THE SOCIAL INSTINCT

THE SOCIAL INSTINCT

How Cooperation Shaped the World

NICHOLA RAIHANI

JONATHAN CAPE
LONDON

1 3 5 7 9 10 8 6 4 2

Jonathan Cape, an imprint of Vintage, is part of the
Penguin Random House group of companies whose addresses can be
found at global.penguinrandomhouse.com.

First published by Jonathan Cape in 2021

penguin.co.uk/vintage

A CIP catalogue record for this book is available from the British Library

ISBN 9781787332041 (hardback)
ISBN 9781787332058 (trade paperback)

Typeset in 11/15.5 pt Mercury Text G1
by Integra Software Services Pvt. Ltd, Pondicherry

Printed and bound in Great Britain by Clays Ltd., Elcograf S.p.A.

The authorised representative in the EEA is Penguin Random House Ireland,
Morrison Chambers, 32 Nassau Street, Dublin DO2 YH68

For my mum, who was a victim of cooperation.

CONTENTS

INTRODUCTION

> I do not think that there is any evidence that man ever existed as a non-social animal.
>
> Charles Darwin, 1871

As I type this, I sit at the kitchen table, practising the social distancing necessary to prevent the spread of a tiny pathogen that has blazed around the globe, changing our lives and routines in ways that seemed unimaginable just a few weeks ago. Workplaces have closed, cafes and restaurants are boarded up, and children are being schooled from home. Over the course of 2020, more than 64 million people would be infected, and 1.5 million killed. As convoys of army trucks lined the streets of small Italian towns to take the dead, and people were forced to say goodbye to their loved ones through phone screens, the rest of the world could only watch helplessly. This virus showed that we are vulnerable to forces of nature, including to things so tiny we can't even see them.

In some ways, the power of a tiny strand of genetic material is extraordinary: how did something that is 500 times smaller than the diameter of a hair destroy so much human life? The answer, of course, is that the virus's success was built on our own. By hunting and trading in wildlife, we enabled the novel coronavirus to jump from one species to the next. It crossed borders on our planes and boats. Like the pathogens of the past, this virus took advantage of the one thing that left humans uniquely exposed to a pandemic: our sociality.

Our social natures got us into this pandemic, but they are also our only way out of it. Although, as I write, it is not yet clear when we will emerge from this threat, we know how it will be done. To confront the virus, we must inhibit our most basic instincts that tell us to seek one another out in times of threat. We must accept constraints on what we can do, with whom and where. To confront the virus, scientists must strive towards a collective goal of finding a vaccine, while society's most essential workers provide the core services and commodities that we all need to survive. To confront the virus, political leaders need to consider not just their own constituents, but people living in other states and even in other countries. To confront the virus, we must *cooperate*.

Fortunately, this is something that we are extraordinarily good at.

*

This is a book about cooperation, the focus of my research since 2004. In a colloquial sense, the word 'cooperation' has become synonymous with bland corporate metaphors, evoking images of firm handshakes and cheerful teamwork. But cooperation is much more than this: it is sewn into the fabric of our lives, from the most mundane of activities, like a morning commute, to our most tremendous achievements, such as sending rockets into space. Cooperation is our species' superpower, the reason that humans managed not just to survive but to *thrive* in almost every habitat on Earth.

Less obviously, cooperation is also the reason we exist in the first place. At the molecular level, cooperation is ubiquitous: every living thing is composed of genes cooperating within genomes. Move up a rung of the ladder and you find the evolution of organisms, where multiple cells work together to make individuals. In most species, the cooperation stops there. Individuals, by and large, don't tend to help other individuals. However, there are a few outliers – and they just so happen to be some of the most extraordinarily successful species on the planet. We are among them.

It is tempting to believe that our hyper-social nature marks us out as different, but the truth is that humans are just one of many species that live a social life. As a particularly poignant example of cooperation, take the Brazilian ant *Forelius pusillus*. By day, these ants forage above ground but, as evening falls, they retreat to the safety of their nest underground. Not all of them, though. A few ants remain outside, waiting for their sisters to scurry down the tiny tunnel. Then they get to work, dragging and carrying grains of sand and other debris to completely conceal the nest entrance from the outside. By sealing the nest, the workers seal their own fate, as solitary ants cannot survive overnight above ground. But dying near the nest could attract predators. In a final stoic gesture, the workers march into the desert night, dutiful protectors to the last.

Ant suicide is an extreme example, but cooperation is so often the key to understanding social behaviour, from the mundane to the awe-inspiring. Cooperation can tell us why parents care for their babies – and why some offspring kill their mothers. It explains why pied flycatchers help their neighbours, whereas chimpanzees try to eliminate them. Cooperation sheds light on things we might never have thought to question before, such as the existence of menopause and sterile grandmothers, and why we are the only primate species that has them.

Of course, we cannot talk about cooperation without acknowledging the flip side: cooperation is infinitely vulnerable to cheats and freeloaders that exploit the collective for their own gain. As we shall see, often these social cheats *are* cooperating, but in ways that impose costs on others: cooperation frequently has victims. Cancer cells cooperate with one another inside multicellular bodies, to the detriment of the sufferers. Nepotism, corruption and bribery are all forms of cooperation, occurring among a minority of individuals while imposing wider costs to society.

Looking long and hard at precisely what humans and other animals do and do not have in common, through the lens of some of our most spectacular and mysterious collective behaviour, has been the mainstay of my academic career. Although I now mostly study humans, I followed a winding path round the globe to get here. I've studied pied babbler birds in the Kalahari Desert, and Damaraland mole-rats in a broom cupboard in Pretoria; apostlebirds in the Australian outback, and cleaner fish living on tropical coral reefs. This may sound like a motley assortment, but these species have something important in common: they all cooperate. In babblers, apostlebirds and mole-rats, cooperation occurs primarily within families. Cleaner fish do things differently, by helping complete strangers – fish they have never met and might never encounter again. And humans are especially interesting because we do both.

*

Cooperation is part of our history. It will also define our future, a fact that has been brought into sharp focus by the 2020 coronavirus pandemic. At the point that the UK went into lockdown, I thought I had finished writing this book. And yet, suddenly, cooperation had never seemed more relevant. The themes I'd woven into these chapters – family ties, community spirit, and policing cheats – became daily headlines on our news channels. The question that I had spent my career thinking about became the puzzle that we now urgently needed to solve: how can we encourage billions of individuals to make a personal sacrifice for a greater good, to prioritise the 'we' over the 'me'?

To tackle such a question requires a broad canvas. We need to look back into our species' evolutionary history to determine how the environments of our past left their mark on the humans of today. But we cannot examine ourselves in isolation: we will also look sideways, to the other social creatures that live on Earth. It is tempting to focus on our closest living relatives – the great apes, and especially chimpanzees and bonobos – but this approach is blinkered. Social behaviours that have the distinct whiff of *humanness* about them are often absent in apes and monkeys but do appear in much more distant connections. Ants and meerkats teach, for example, but chimps don't. Pinyon jays will share their stuff, but bonobos won't. The clues to understanding our own place on Earth exist not only in comparisons with other primates but on totally different branches of the evolutionary tree, where we find species who live, like us, together.

This book is structured in four parts, and I think of them like the layers of an onion. In each part, we will move from the centre to the next outer layer, exploring the evolution of social complexity at different scales as we go.

Part 1 explores the evolution of individuals. We are going to start small, delving deep inside ourselves to witness how genes and cells work together to form coherent entities: you and me and every other living thing. Our bodies appear to be unified fronts, but conflict lurks beneath the surface: genetic and cellular cheats frequently try to subvert the social order, sparking inner chaos when they do. As we shall see, our health, fertility and even our survival can depend on our ability to hold these selfish entities in check.

Part 2 zooms out a little, exploring the evolution of families. Caring for our children seems natural to us, but we are unusual in the extent and duration of parental investment we provide, as well as the fact that this care often comes from fathers as well as mothers. We will discover what we have in common with the other species that live in extended family groups, and how our familial heritage can explain many of our strangest features, including why women experience the menopause and our extraordinarily long lifespans.

In Part 3, we will widen the net, asking why help is sometimes extended beyond the family circle. Part of what makes us human is our ability to cooperate not just with relatives and friends, but with complete strangers and people we might never see again. It is these tendencies that underpin our hyper-social natures, and that paved the way for humans to colonise the globe. But we aren't the only ones that socialise with strangers: here, we will meet the cleaner wrasse, a small reef-dwelling fish with which we have a surprising amount in common.

In Part 4, we will jump to the outermost layer of our onion, to explore the evolution of large-scale societies. Here, we will acknowledge our ape heritage, asking how and why we came to be so different. Interdependence shaped our psychology, allowing us to achieve ever-greater feats of cooperation, but simultaneously left us exposed to the vagaries of social comparison and paranoia, that can become pathological for a minority. Our supreme commitment to cooperation is the key to solving the massive global problems we now face, but it is our ability to cooperate that might also be our eventual downfall.

We're going on a journey to explore the evolution of cooperation, a journey that will tell us more about ourselves and the other species with whom we share this planet. We will discover that cooperation didn't just change the world, it *made* it. Every human achievement we can think of, from the trifling to the truly magnificent, relies on cooperation. What's more, without it, there would be no life on Earth at all. In the first part of this book, we are going to discover how cooperation created you and me and every other living creature. Let's dive in.

Part 1

THE MAKING OF YOU AND ME

Cooperation is the crucial ingredient in cooking up life as we know it. Without it, neither you nor I, nor any other living organism, would exist. You are constructed of genes cooperating within genomes within cells, all working together for the greater good of making you *you*.

To illustrate this point, think of yourself (or, indeed, any other living thing) as a Russian doll. You are you on the outside, but this external appearance isn't all there is. If you prise the outer shell open, you'll find another version of you lurking inside, which itself contains another version and another within. You are simultaneously an individual and a collective. Your body is made up of trillions of cells – around 37.2 trillion to be precise. For perspective, that's more than 5,000 times the number of people on Earth. Most cell types contain forty-six chromosomes (sex cells have twenty-three – and red blood cells have none) and each of these accommodates genes, ranging in number from a few hundred to many thousands. If what you see when you look in the mirror is the outermost Russian doll, the gene is the smallest one, housed deepest within. This innermost doll is the most fundamental of them all – the one that cannot be split any further. It can persist long after your chromosomes and cells – and you – expire. In your body and mine, the gene is hidden away. To make its way into subsequent generations, the gene must travel with the outermost doll – the vehicle of reproduction.

But life doesn't have to be this way. The units you are made of are, at least in principle, capable of independent reproduction: genes don't always have to be encased in cells, and cells *can* reproduce even if they are not stuck inside bodies. To climb the ladder of complexity, from genes to cells to organisms to groups, requires the units on the lower rungs to rein in their own self-interest: they must cooperate.

The truly pivotal moments in our history were not the invention of the wheel or the signing of the Magna Carta or the domestication of crops and animals, but rather a handful of fleeting events that occurred long ago in Earth's history. To gain a sense of just how improbable our multicellular existence is, it helps to take a historical perspective. We need to go back as far in time as we can imagine, to the origins of the solar system and our planet.

*

Earth is estimated to be approximately 4.5 billion years old. This is an impossibly large number to grapple with – so instead let's think of Earth's history scaled down to a single calendar year. Our species has only just entered the stage, appearing late into the night on 31 December, just half an hour before the present day. Once we had arrived, it took us a mere twenty-five minutes to spread around the globe. Within the last sixty seconds, we have engineered both the Agricultural and Industrial Revolutions, formed global nation states, waged two horrifying world wars, and presided over wholesale domination and destruction of the natural world.

With all that has happened in the last thirty minutes, it is astonishing to learn that the first life on Earth – genes housed within cells – appeared way, way back in mid-March. In fact, the shift from independently replicating strands of genetic material to bona fide cells is thought to have happened just once in the entire history of our planet. Every cell in every single organism that has ever lived is a descendant of this prototype. By June, the first eukaryotic cell had appeared. These are the more sophisticated cells that appear in all complex life forms, including plants, fungi and animals. Again, as far as we know, this transition happened just once. Around November, cells made the leap from being sole traders to forming confederations with other cells, and multicellularity was born. These shifts were evolutionary milestones, crucial steps along the road that took us from free-living strands of genetic material to the emergence of life in all its shapes and forms.

A common theme permeates the major evolutionary transitions: each one involved smaller units locking themselves away inside a larger one – the newly-created outermost shell of the Russian Doll becoming a new level of biological organisation, a new kind of 'individual'. This expansive view of life on Earth therefore reveals a history of partnerships, of singletons becoming teams and working towards collective goals. The history of life is a history of cooperation.

Very occasionally, these partnerships can take on staggering proportions. The largest collective on Earth, for example, is a six-legged beast, with its feet in North America, Asia, Europe, Australia, Hawaii and New Zealand. This colossus is a single super-colony of a highly invasive species, called the Argentine ant. Most ants live in nests, comprising a single breeding female and hundreds or thousands of workers. Within nests, interests are (more or less) aligned: workers collaborate to raise the offspring produced by the queen. Between nests, however, competition is fierce: workers fight, often to the death, when they encounter rivals from a different group.

But Argentine ants seem to be different: the workers from different nests seem to see one another as allies rather than enemies. Taking an ant from a nest in New Zealand and introducing it to an ant from Italy prompts nothing more than the ant equivalent of a passing nod. This relaxed attitude is not due to the Argentine ant being unusually docile: on the contrary, these creatures will unceremoniously attack and kill ants that they do not recognise. It turns out that the colonies in Italy and New Zealand are part of a massive, international super-colony, with millions of nests, spanning different countries and continents, being descended from a single mother unit. Introducing an ant from New Zealand to an Italian counterpart therefore appears to be the metaphorical equivalent of you touching your big toe with your little finger – both body parts belong to the same larger unit. Although relations between members of the same super-colony are friendly, competition between ants from different families is ferocious. In California, the long, narrow border between two different super-colonies is an ant war zone, littered with the dead bodies of the soldiers from these two rival camps.

It is easy to be impressed by the scale of cooperation in the global super-colonies of Argentine ants. But cooperation at any level of life is something we should marvel at. Joining forces – even fleetingly – entails personal sacrifice for a greater good. Acknowledging this fact plants a central question in the ground, in territory where the natural and social sciences meet. How can we reconcile personal sacrifice with a Darwinian view of evolution, with its emphasis on self-interested individuals?

Before we can answer this, it will be helpful to cover some of the basic concepts about how evolution works. Travel back in (not so distant) time with me, to a dusty Cambridge college, and a moment that almost ended my scientific career before it had begun.

1
A COLD SHUDDER

> You can live some sort of life and die without ever hearing the name of Darwin. But if, before you die, you want to understand why you lived in the first place, Darwinism is the one subject that you must study.
>
> Richard Dawkins, foreword to John Maynard Smith's *The Theory of Evolution*, 1993

In October 2000, I was embarking upon a natural sciences degree at Cambridge University, not quite sure what I had let myself in for. As was usual, four of us had squeezed into a small office to have our weekly meeting with one of the college tutors, where we would discuss the course material and make sure that we understood the content. Our tutor, Veronica, was a botanist with a bashful manner. She spoke in a quiet voice, barely more than a whisper, and offered her remarks on our written work in light pencil strokes as if afraid of stating her case too forcefully. At the end of an hour's discussion about something 'planty', the details of which I cannot recall, Veronica announced that she wanted us to go away and write a 2,000-word essay on 'why contemplating the eye gave Darwin a cold shudder'.

I panicked. I had picked the 'Evolution and Behaviour' course for the 'Behaviour' part, believing (mistakenly as it turned out) that this lay within the purview of psychology, not biology. I hadn't the faintest idea about evolution and hadn't read any Darwin. How was I supposed to fathom what might bring him out in a cold sweat? In those days, it was not as easy to rely on Google or Wikipedia to glean any clues as to what the hidden meaning behind this obscure question might be. I started to wonder if coming to Cambridge had all been a big mistake. If I couldn't even understand the question, how on earth was I going to come up with an answer? I don't remember what I wrote for the essay but I received a decidedly average mark for it.

Though I was wholly unprepared to answer it, the cold-shudder question cuts to the heart of the difficult problem of explaining the appearance of complex design in nature. The human eye is a marvellously intricate organ. It has a lens, allowing us to focus on objects whether they are far away or right under our nose. Along with other mammals, we have colour vision, meaning that we can see somewhere in the region of 100,000 to 10 million different colours. We have different photoreceptor cells, called cones and rods, that are specialised for their roles in daytime and night vision. This is not a book about the eye and I don't wish to labour the point, but the eye is a pretty amazing bit of sensory kit. From an evolutionary perspective, however, the eye is a bit of a puzzle because the Darwinian mantra is that complex adaptations derive from a series of small, gradual steps, each one bringing a slight improvement in performance and advantage to the bearer. If an eye is only functional once it can actually see things, what possible use is a half-formed version?

Darwin himself acknowledged this difficulty with his grand theory. In response to a colleague's gentle criticism of his book, Darwin confessed, 'About weak points I agree. The eye to this day gives me a cold shudder.' Even now, proponents of creationism brandish the eye as incontrovertible proof that Darwin's theory must be wrong, and that the appearance of design implies an intelligent omniscient designer. Perhaps this was why he worried about it.

But did he, though?

While Darwin admitted that viewing the eye as being the product of natural selection seemed 'absurd in the highest possible degree', he went on to speculate how a complex eye could, in principle, evolve in successive steps, starting simple and becoming increasingly sophisticated. For this to happen, the tiny changes would have to be passed from parents to offspring and be beneficial to individuals who inherited them. Although it was not until 150 years later that he was vindicated, Darwin's hunch turned out to be prescient. We now know that complex eyes did evolve in a series of gradual steps – starting out as simple layers of photosensitive cells that allowed individuals to regulate their daily cycles, and incrementally layering on successive features that, at each step, proved advantageous to the bearer.

*

Variation in traits is most commonly passed from parents to offspring via genes: packages of information that are transmitted, unchanged, from one generation to the next. Genes contain the instructions that your cells need to build proteins; it is these proteins that are the workhorses of life. Your bones, your skin, your fingernails and your hair are all made of protein. And so is your brain. Your thoughts, feelings and moods are all events taking place in structures made of protein.

Evolution is change over time – in a biological sense it refers to the rise and fall of different gene variants* in a population. This ebb and flow of gene variants can occur because of natural processes. Mutations introduce new variants into populations, while stochastic events, like asteroid strikes or volcanic eruptions, can wipe out an entire gene lineage purely by chance. But there is just one force

* A gene variant (also known as an allele) is the particular version of a given gene. For example, everybody has genes that determine eye colour, but it is the particular variants of the genes that determine what colour your eyes are.

that consistently *pushes* gene variants in one direction or another: we call it natural selection. This is the process by which gene frequencies change *because of the effects the genes have on the bearer*. Selection is a blind sorting mechanism: when there is variation in traits, and variation can be inherited by offspring, gene variants that code for beneficial traits will tend to accumulate in the population. As Darwin emphasised, these differences could be slight: 'a grain in the balance' yielding just the 'slightest advantage' would be sufficient to drive this great engine of change.

All else being equal, a gene variant will tend to increase in frequency in a population when it codes for a trait that confers a survival or reproductive advantage on its bearer, allowing it to outcompete individuals who don't carry that gene variant. Variants which are advantageous to their bearers – which affect physical or cognitive traits that either increase survival or reproductive success – will tend to accumulate in the population. To use the language of evolutionary biology, these genes will be under positive selection. When we seek to explain why certain behavioural traits, such as aggression or caring for offspring or being kind to strangers, exist and persist within populations, we are implicitly asking how the genetic variants associated with those traits are favoured by selection. This doesn't imply that behaviour is either exclusively or deterministically governed by genes, or that genes exert the same effects in all the bodies or environments that they find themselves in. Nevertheless, for traits that *do* have some genetic component, however small, we can ask how likely it is that these genes will find their way into subsequent generations based on the effects they help to produce.

Taking the gene's perspective is sometimes called the 'gene's-eye view', most famously championed by Richard Dawkins in his treatise of 'selfish' genes. Genes *are* selfish but it is important to clarify what this rather loaded term really means. Describing genes as selfish is not implying that they are immoral or conniving or any of the other unsavoury character traits we might attribute to selfish *people*. It is also not a shorthand for describing genes that tend to be associated with selfish characteristics, residing only in the bodies of the most nefarious individuals. On the contrary, every one of the approximately 25,000 genes in your body can be described as 'selfish' or, less controversially, as 'self-interested'. Genes are self-interested in the sense that they each have a singular overriding 'concern'*: to ensure that they appear in subsequent generations.

Taken at face value, the selfish-gene perspective seems to imply that any heritable trait that lowers an individual's reproductive success or survival (and the gene variants which underpin it) will be ruthlessly weeded out of the population. But if we accept this world view – with its narrow reading of Darwinian logic – then how do we account for the many examples of cooperation we see in the world around us?

As a concrete example of the phenomena we are trying to understand, think back to the case of the suicidal ants we encountered earlier. At first glance, the existence of such extreme altruism seems to pose a serious challenge to Darwin's theory. Darwinian logic hinges on the assumption that individuals are driven by self-interest. Most creatures try to survive and have as many offspring as they can, even if this means that there are more hungry mouths than the environment can support. Natural selection acts as a hidden sorting mechanism, a metaphorical sieve: when there is not enough to go around, then only the strongest, the fastest, the *fittest*† survive. How and why would evolution have promoted these heroic

---

* Of course, genes don't really 'care' about anything – they are just scraps of genetic material, incapable of experiencing wants, needs or desires. When we say that genes 'want' or 'care' about something, what we really mean is that genes act *as if* they care about their survival and *as if* they want to make it into the next generation.

† In everyday language, we use the word 'fitness' to refer to athletic prowess but in evolutionary biology this word has a different meaning. It refers to an individual's (or, more accurately, a gene's) average contribution to the gene pool in the next generation. Fitness is therefore a measure of evolutionary success. When I use the word fitness, it is in this evolutionary sense.

tendencies, that impose the gravest costs on their bearers while directing benefits to others?

The key to understanding the curious ant behaviour is to appreciate that members of the same colony are highly related. It is no coincidence that some of the most striking examples of cooperation in nature occur within the confines of family groups. To understand why charity so often begins at home, we need to expand our view of the ways in which an individual's actions might yield downstream benefits to his genes – and we can do this by taking the gene's-eye view. A gene that is present in both my and my brother's body doesn't care how it gets into the next generation. From the gene's perspective, it is irrelevant whether it is transmitted via my own children or via my nieces and nephews. Costly helping behaviours can be favoured by selection if the benefits to relatives (in terms of increased offspring) sufficiently compensate the costs (in terms of foregone reproduction) that the helpful individuals face.

This overarching framework for explaining the evolution of social traits is called 'inclusive fitness theory' and its logic allows us to make specific predictions about where helping behaviour will evolve, and to whom it will be directed. The ants that sacrifice their lives to seal the colony from outside don't incur a personal reproductive cost because ant workers are sterile. Moreover, scores of relatives stand to benefit when they perform this selfless act, which can explain how such extreme sacrifice could be favoured. Relationships among siblings in human families can also be affected by relatedness: research shows that full siblings tend to see more of one another and invest more in the relationship than they do with half-siblings, even when half-siblings are raised under the same roof like full siblings.

*

Relatedness is a crucial factor in explaining why individuals help one another, and this expanded notion of self-interest allows us to reconcile otherwise puzzling behaviours with Darwin's theory. But relatedness can't do all the work. The benefits and costs must stack up as well. Benefits and costs are ecological parameters that also dictate the circumstances under which cooperation will be favoured. One way to give cooperation a shove is to reduce the costs associated with helping others – this can happen when it is difficult for individuals to breed independently.

A particularly charming example is that of the long-tailed tit. These gregarious birds are seasonal breeders who flit about in large, chattering flocks but usually try to breed independently, in pairs. Despite their dainty size, long-tailed tits build the most elaborate nests of any European bird species. The nest is a completely covered dome, with a small cavity at the mouth for the parents to pass in and out, and is constructed of materials that seem to have been drawn from a witch's spell. The exterior calls for scraps of moss and lichen, bound together with spiders' silk; while the interior is lavishly furnished with thousands of plush feathers. The tits spend over three weeks constructing these fairy-tale palaces, a sizeable investment in the context of the barely three-month-long breeding season. Nevertheless, for many birds, the effort is in vain: most nests are discovered by predators, with fewer than one in five pairs successfully fledging their young. If their nest fails when it is close to the end of the breeding season, the hapless pair will typically seek out relatives nearby and help at that nest instead. In this example, the cost of helping is very small: if there is no time left in the breeding season to try for another independent attempt, then the reproductive sacrifice is minimal. These small costs make it relatively easy to tip the balance towards helping relatives instead.

When we look at evolution this way, we can start to make more nuanced predictions about when we should expect help to arise and be favoured by selection. For example, individuals should be less willing to help old or extremely sick relatives, since there is a smaller chance of the investment being converted into fitness benefits. This rather unromantic calculation seems to be performed by African *Megaponera* ants when they raid termite nests. Termite colonies are defended by specialised soldier morphs, who aggressively attack the marauding ants, sometimes wounding them in the process. Injured ants release a pheromone – a chemical cry for help – that stimulates their comrades to rescue them from the jaws of a termite soldier and carry them back to the nest. At the nest, the stricken ants are tended by their sisters, whose licking and grooming prevents the wounds from becoming infected. There's a twist however: ants that are either too badly injured (for example, having lost five legs rather than one) or too old are not rescued. There is no evolutionary advantage to rescuing a nest-mate who is going to die soon anyway. Ants themselves seem to know when they aren't worth the trouble and are less likely to send a distress signal if they are aged or too badly injured.

A study of US households in 1910 reveals something similar might occur in humans. In this database, childless couples were more likely to help their relatives (just like the long-tailed tits I described above). Following the logic that not all relatives are equally valuable, these childless couples were more likely to take a niece or nephew into their home, than an ageing parent. In relatedness terms the parents should be more valuable but, in terms of the potential fitness benefits, ageing parents are an evolutionary lost cause: there is more potential to reap a benefit from helping a young niece or nephew instead.

*

So far, we've seen that selection can favour genes associated with cooperative and even heroic actions when this benefits copies of these genes that are found inside other individuals. Genes may be selfish but this doesn't preclude cooperation, under the right circumstances. Although taking the gene's-eye view can help to explain many puzzling phenomena, you may have noted a problem with the story I have told. Here's the difficulty: gene variants are commonly and rightfully viewed as the entities that natural selection works upon, and whose frequency either increases or decreases over evolutionary time. But genes are bundled up into collectives – which we call organisms or individuals – and it is the effects that genes have on these individuals that are exposed to selection. To put it another way, genes do not bear the adaptations that determine their own success. Instead, these design features are carried at a higher level of biological organisation, by the individual.

The invention of individuals was an evolutionary masterstroke. Take a moment to consider what a multicellular individual really is: you and I and every other multicellular being on Earth is a *collective* that operates as a whole rather than as a bunch of parts.

Earlier I said we could think of genes as if they were tiny agents, pursuing their own agenda. But the individuals we see all around us seem to be goal-driven in a similar way. An individual oak tree reaches towards the sun *as if* its goal is to grow ever taller; an individual great tit carries food to the nest *as if* its goal is to help the young chicks grow and survive. Behavioural ecologists, like me, tend to talk of individuals, rather than the genes, as being the entities that pursue an evolutionary agenda because it is individuals, rather than genes, that we can see and whose behaviour we observe.

But this mental shortcut is justified. Evolution creates individuals by aligning the interests of the genes within them. An individual that pursues her own evolutionary agenda is therefore pursuing the agenda of all the genes from which she is made. This equivalence allows us to think of individuals as being goal-driven agents, safe in the knowledge that we can translate back to the gene's-eye view at any point.

The evolution of individuals was the first crucial step on the road to ever-increasing social complexity, foreshadowing the transitions to families, to communities and to large-scale societies. But how do we know when a collection of genes and cells becomes an individual? Why should the person reading this book have the special status of being the individual, rather than conceiving of every cell in your body as an individual or, indeed, every selfish gene?

It feels intuitive to conceive of ourselves as coherent individuals, while simultaneously denying that status to a flock of seagulls or to a herd of wildebeest. But appeals to intuition are an unreliable tool for demarcating these evolutionary boundaries. For example, it probably feels rather *counter*-intuitive to conceive of a colony of ants as an individual, but many evolutionary biologists would argue that ant colonies are super-organisms in their own right. On what basis, other than intuition, can we draw the line? Are you *really* an individual and, if so, why?

2
INVENTING THE INDIVIDUAL

> The individual is, accordingly, a unified commonwealth in which all parts work together for a common end.
>
> Rudolf Virchow, *Atoms and Individuals*, 1859

Evolution invents new individuals by sewing the interests of the parts tightly together with the whole. Remember the Russian-doll analogy, with the inner dolls representing genes and genomes and cells. The inner dolls have only one route to the next generation: they must travel inside the outer doll, the 'individual'. This imprisonment aligns the interests of the inner dolls, meaning that they are incentivised to work together, rather than against one another. Their joint mission is to create the best individual they possibly can, with their ticket to the next generation depending more or less entirely on their success in this collective venture.

Individuals can exist as single cells, as multicellular organisms, or even as entire colonies in some cases. To distinguish individuals from groups or collectives, it helps to recognise that natural selection is not just a process, but an engineer: a force that assembles an entirely new product from a collection of parts. Human engineers frequently have design goals in mind: the people designing a new iPhone will have a target size and weight in mind, as well as other specifications for the quality of the camera and the battery life, for example. Evolution has no foresight, but nevertheless shapes and moulds the design features of individuals in a similar way, by sifting out the less fit (or worse adapted) variants from the population. We can identify *individuals* by spotting the level at which the design features seem to cohere, and to pull in the same direction.

This is a bit abstract so, to illustrate, let's use a real-life engineering example: the car. The buyer of the car is the hand of selection, and the car is the individual. Just like a multicellular organism, a car is made up of functional subunits. For example, the crankshaft, spark plugs and pistons that comprise the engine are analogous to the cells, nerves and muscles that form organs like the heart in your body. These are *sub*units rather than units in their own right because their functionality is only obvious when considered as part of the car. Assuming that you had no prior knowledge, if you were to find a steering wheel or a piston or even a full engine abandoned on the street absent the vehicle, it would be difficult to determine what these parts were designed to do, whereas their role becomes obvious when they are working in synchrony with the car's other parts. An individual therefore has the appearance of being goal-driven and possesses functionality in the same way that a car does – but that the component parts lack.

What's more, when working together, the subunits of the car give rise to features – or adaptations – that are only apparent as properties of the car itself and not of the constituent parts. These design features are what we consider when selecting which car to buy. We don't pore over the sump and the fan belt – instead, we select based on the properties and features of the car itself. Does it look good? How fast can it go? Is it reliable? By selecting on these properties, we are of course exerting a selection pressure on the component parts despite the fact that we don't see (or really care) about them. If speed or reliability are things that consumers care about to the extent that they affect differential car sales, then this selection pressure will drive changes in the componentry, in that some kinds of pistons and crankshafts will be more likely to end up in car engines than others. Natural selection sifts among genetic variants in a population in much the same way, by acting upon the design features of the individuals that carry these genes rather than directly upon the genes themselves.

*

Such an analogy helps to make the case for conceiving of highly social insect colonies (for example ants and termites) as being individuals in their own right – or 'super-organisms'. Social insect colonies often exhibit striking similarities with multicellular bodies, like yours and mine. In particular, the design features and behaviours of the constituent insect 'parts' can only be understood with reference to the higher level of organisation: the colony. Many social insect colonies are headed by a single queen – the only truly reproductive member of the colony. She is akin to the egg-producing ovary in a multicellular body. The workers are the queen's sterile daughters, who help her to produce more offspring. These workers are equivalent to the non-reproductive ('somatic') cells in your body, that generate the body parts and do the majority of the maintenance and repair work needed to keep things running smoothly. The sterile cells in your body and the sterile workers in an insect society can only be understood with reference to their roles in the higher unit: the body or the colony. A sterile insect without a colony is an evolutionary mistake but inside a colony it is an evolutionary miracle.

The parallels don't end here. Just as body parts are specialised for different roles, social insect workers vary widely in their appearance and the tasks they perform around the colony. Some develop into soldiers, with a large body and biting mandibles, while others focus on foraging or looking after the developing brood. Division of labour is taken to the extreme in the turtle ant, a species that nests inside the dead branches of red mangrove trees in the neotropics. In this species, some workers develop large, disc-shaped heads, becoming specialised as living doors that plug the holes in the branches that might otherwise leave the colony vulnerable to intruders.

Our bodies have immune systems that detect signs of cellular damage and the presence of pathogens. Cells that show evidence of mutation or decay will undergo a highly controlled form of cell death, called apoptosis, whereby the cell issues itself an instruction to self-destruct. Infected ants seem to do something remarkably similar. *Temnothorax* ants who are infected (either naturally or experimentally) with a fatal and infectious fungus act as if they know that they are a danger to the colony. An infected ant ceases all social contact with her sisters, and nobly removes herself from the colony to die in isolation.* Ants that care for the developing pupae can also detect when the young are infected with these kinds of pathogens, and will selectively terminate young that show signs of infection.

Insect colonies can also regulate their core temperatures, much like birds and mammals do. As a warm-blooded mammal, you are able to maintain your core body temperature within a narrow range: from 36.5 to 37.2 degrees Celsius. Temperatures outside of this range can quickly become dangerous and potentially fatal. We have evolved various mechanisms for regulating body temperature: drinking water and sweating can cool us down, as can sending blood to the surface of the skin (resulting in a flushed appearance when hot). When we need to warm up, the hairs on the surface of our skin stand on end, trapping a layer of warmer air close to the body and, in extreme cases, our muscles may begin to involuntarily contract as we shiver. Astonishingly, social insect colonies also thermoregulate in this way. The best studied case is the honeybee, where, on hot days, some workers are tasked with the role of bringing water to the

* A rather extreme example of the 'social distancing' that is necessary to slow the spread of a transmissible infection.

hive and spraying it over the combs, allowing evaporation to cool the hive from within: the colony sweats! In the winter, when the hive can become too cold, other workers act like the shivering muscles in your body, by vibrating their flight muscles rapidly to generate heat. To power this energetically demanding activity, other bees in the colony shuttle honey back and forth to fuel the heater bees. In both multicellular bodies and in social insect colonies, therefore, the constituent parts seem to bear adaptations that are designed to promote a greater good.

*

We might be able to conceive of social insect colonies as super-organisms, but what about human groups? Are our societies also super-organismal and are we, the people living in them, akin to the worker bees and ants that toil for the benefit of their colony? Some evolutionary biologists believe that the answer to these questions is yes. Like the insects we just met, humans also have widespread division of labour and are tremendously cooperative, including in scenarios where help is not directed to kin and we can expect no return favours from the beneficiary. These evolutionary biologists claim that our species' uniquely cooperative nature only makes sense if we consider ourselves as being cogs in a larger machine. So the argument goes, cooperation can only be understood because of the benefits this yields at the group level, with the implication being that selection also operates at this higher level of biological organisation. While I concede the point for the social insects, I don't agree with the proposition that human groups are super-organisms.

Before I explain why not, we should remind ourselves of what is involved in creating a new kind of individual. For a collection of parts to be welded into a new kind of being, their interests need to be almost completely and permanently aligned. When this happens, then the component parts will surrender their personal autonomy, and work together to produce a greater good. The easiest way for this to be achieved is for the constituent parts to be highly related to one another because high relatedness tends to suppress the scope for conflict. In the most extreme case, if you were to gather a group of clones together, they would 'care' as much about one another as they do about themselves and conflict would be absent. Of course, this is precisely the situation the cells in your body find themselves in: you developed from a single cell, and most of the cells in your body are clones of this progenitor. As a consequence, in multicellular organisms, selection (usually) acts at the individual level and generates adaptations that are expressed at this new higher level of organisation.

Human societies are not formed of clones and, unlike social insect colonies, are not giant families either. Nevertheless, there are other factors that can align the interests of group members; one prominent way being when there is a threat from a rival group. The television programme *The Apprentice* neatly illustrates how competition between groups can foster cooperation within them. The show centres around a cohort of hopeful contestants that compete for the right to be hired by the irascible businessman, Lord Alan Sugar. Each week, contestants are assigned to one of two groups to undertake a business challenge – designing and marketing a new chocolate bar, for example – with the winning team being the one that makes the most sales. At the end of each week, the most incompetent (or annoying) contestant in the losing team is fired and thereby eliminated from the competition.

When the two teams are in competition with one another, everyone has a common goal to help their team win. This is because no one in the winning team risks being fired. Cooperative teams – where members are helpful and work together – are generally more likely to succeed than teams where members compete against one another. In the laboratory, we can recreate similar scenarios to show that cooperation within groups tends to thrive when we create competition between them. Even children as young as four years old invest more in their team when there is another group to compete against.

But there is a fundamental difference between groups that are formed on the basis of high relatedness, and those where members' interests only align because of happenstance. Inter-group rivalry can prompt teammates to unite in the short term, but these collectives are (more often than not) tenuous and temporary truces. On *The Apprentice*, goodwill rapidly evaporates when contestants find themselves in the losing team. When there is no rival team to unite against, a contestant's tenure in the competition depends on their ability to outmanoeuvre their teammates. A familiar pattern usually results: people swiftly turn on one another, erstwhile allies becoming vicious rivals. Insults fly around the room, as contestants try to absolve themselves of blame while incriminating their useless colleagues. When between-group competition is no longer relevant, then competition within the group becomes much more apparent.

Contestants on *The Apprentice* only cooperate when they face competition from a rival group – and it therefore makes no real sense to think of the teams as being higher-order 'individuals' that maximise collective success. Such a perspective tells us nothing about why contestants stab their teammates in the back when the chips are down. Instead, it makes more sense to think of each team as being composed of several self-interested individuals, who invest in team success when it is to their advantage to do so, but not otherwise. This example also highlights the difficulty with conceiving of human groups as super-organisms, a hypothesis that hinges on the assumption that competition between human groups is (or was) consistently stronger than the competition within them. This assumption must be met if the long-term fitness interests of the individuals and their groups are to be completely and permanently aligned. I don't think that the data is very convincing on this point. Although the interests of individuals do frequently coincide with those of their group, they will also frequently diverge.

To see how individuals can *seem* to act in the interests of the group, but still be understood as pursuing their own individual success, let's use warfare as an example. Our species' willingness to go to war is a poster child for the idea that human cooperation evolves because of group-level benefits. Waging war might not sound prosocial, but in some ways, it can be: individuals sacrifice everything – potentially even their lives – ostensibly for the greater good. It is clear that contributions to warfare can and do yield significant benefits at a group level. But it requires another leap of logic to assert that cooperation in this realm is a group-level adaptation. Instead, decades of anthropological studies conducted in contemporary tribal societies show that contributions to warfare can be beneficial to individuals (particularly to young men, who might use warfare to steal valuable resources, to kidnap women, or to gain warrior status), and can sometimes actually be costly at a group level. The point here is that participation in inter-group competition does not invariably benefit the group: much of the time, these actions are equally consistent with individuals pursuing their own personal interests, with any group-level benefits or costs of the conflict being an incidental by-product.

*

Over the last decade or so, the line between individuals and collectives has become blurred once more by the realisation that multicellular organisms host vibrant and diverse communities of microbial partners. The average human has as many microbial cells in their gut as there are cells in the entire body. Although we tend to think of microbes as being pathogens, many provide significant benefits – to the extent that hosts frequently and actively seek them out. Stink bugs, for example, are so reliant on gut microbes for digestion that their mother deposits a little faecal pellet loaded with these symbiotic bacteria under the eggs, providing the first vital meal for her newly hatched offspring.* Offspring that do not receive these symbiont snacks are unable to digest the plant material that they feed upon. Similarly, the termite's rare ability to digest wood is dependent on a suite of gut micro-organisms. Termites obtain these by eating the faeces and drinking a symbiont-laden fluid from the anus of other colony members.

Although these habits sound unsavoury, it turns out that human infants acquire their own intestinal flora in a not-too-dissimilar way. In utero, infants develop in a bacteria-free environment. An infant's first encounter with microbes occurs at birth: during a vaginal delivery, the newborn intestine is colonised with several species of bacteria from the mother's gut, an event which kick-starts the development of the child's immune system. Children that are delivered by Caesarean section are not exposed to the same microbial community as those who are born vaginally – and receive fewer beneficial gut microbes. One recent study found that the differences in gut microflora between children delivered vaginally or by Caesarean section were still apparent even at seven years of age; and the reduction in microbial diversity that is associated with Caesarean section delivery has also been implicated in increased susceptibility to a range of atopic conditions, such as asthma, allergies and eczema.

Based on the fact that microbes and hosts seem to live highly interdependent lives, some scientists say that microbial and host

* Stink bug is a generic term for the family of insects known as Pentatomidae. Different species of stink bug can be found all over the world, including a particularly invasive species called the brown marmorated stink bug. I recently discovered one of these in my house and – stupidly – caught it in my hands. If you make the same mistake, then bicarbonate of soda is the only way to remove the stench. Stink bugs are also common in the Kalahari and I've lost my dinner more than once due to a stink bug spaffing in my meal.

genomes ought to be seen as a single coherent unit – called the hologenome – and that hosts and microbes exist as a single indivisible entity: the holobiont.

I don't think such a sweeping category makes sense, though. Many micro-organisms upon which multicellular hosts depend are acquired from free-living populations in the environment, showing that the fortunes of microbes and their hosts are not inextricably intertwined. Each host is home to hundreds or even thousands of different microbial species: it is unlikely that these many different assemblages form a single coherent unit of selection. Microbes that are ordinarily harmless or even beneficial to us can switch allegiance inside our bodies, becoming dysregulated and even aiding the growth and proliferation of cancerous cells. In short, microbes may provide benefits to hosts but that is a far cry from saying that these traits have evolved *because* of the benefits they confer. The microbiome is an important partner – but it is not part of you, the individual.

Nevertheless, there is one bacterial partnership that all multicellular beings have which *is* different – and does have the permission to consider itself a part of the individual. This is the mitochondria – the internal battery that lives within each cell in your body. Gaining these little power sources was the key innovation in the formation of the eukaryotic cell, an event that happened just once in the history of life on Earth and which gave rise to the multicellular branches on the tree of life, including all animals and plants.

Mitochondria all but certainly existed as free-living bacterial organisms that, by some quirk of evolutionary luck, found themselves ensconced within another cell. The specifics of exactly how this occurred are unclear – though one plausible scenario is that the host cell was a phagocyte (a single-celled predator) that engulfed the smaller mitochondria without digesting it. This cellular indigestion proved advantageous both for host and captor: the captor received a free energy supply while the entrapped mitochondria gained protection from the outside world.

The energy supplied by mitochondria meant that eukaryotes could grow much larger (15,000 times on average) than prokaryotic cells, potentially allowing eukaryotes to exploit new ecological niches and to feed on other, smaller cells by simply engulfing them. The extra energy also meant that the eukaryotic cell could have a higher metabolism: in other words, it could do more stuff. Most of a cell's energy budget is spent on protein synthesis, so having an onboard battery allowed eukaryotes to increase their genome size and to synthesise proteins at higher rates. By spewing out proteins and combining them in different ways, a cell can grow in size and complexity, creating different internal departments and allocating different functions to them.

Despite their bacterial ancestry, mitochondria are part of the whole in a way that a microbial community is not. The reason is that the mitochondria have permanently committed themselves to the symbiont life – their only means of reproduction is to travel with the cells that make up the organism. This re-emphasises why individuals are truly invented by evolution: making one requires the almost total suppression of conflict between the units that comprise it.

Almost being the operative word.

3
THE RENEGADES WITHIN

Mother Nature is a wicked old witch.

George C. Williams, *Evolutionary Ethics*, 1993

So far, we've seen cooperation as a creative force, one that can package genes and cells together to invent new beings. But wherever cooperation exists, we know that there is the potential for conflict. The cells that make up our bodies occasionally squabble and bicker, and similar conflicts also occur between the different genes residing inside the cells themselves. For cooperation to prevail, and for the integrity of the whole to be preserved, these inner conflicts must be reduced or resolved.

Remember that our cells contain smaller subunits, called mitochondria, and that these formerly free organelles opted to join a cellular team rather than going it alone. I said in the previous chapter that there is negligible conflict between mitochondria and the host cells, but that doesn't mean that the marriage is entirely harmonious. Unlike the nuclear genes, the handful of genes that reside in the mitochondria are only passed down the female line. All of the mitochondria in your body came from your mother, just like all hers came from *her* mother, and so on. Genes which reside on the mitochondrial genome therefore favour female offspring because ending up inside a male body is an evolutionary dead end.

This intragenomic favouritism can lead to stark differences in the quality of male and female offspring, a phenomenon known as Mother's Curse. Mitochondrial genes that confer an advantage to female offspring can be retained, even when these same genes have severe damaging consequences for males. For example, Leber's hereditary optical neuropathy is a disease that typically appears in adolescent males and is caused by mitochondrial genes. Sufferers experience the degeneration of the optic nerve and, ultimately, bilateral loss of vision.

In Canada, all male sufferers of this disease have been traced back to a single female – a woman who was sent over to what was then called Little France in the late 1600s. Little France had too many men, and women were sent to the colony by Louis XIV in an effort to rebalance the highly skewed sex ratio and help found a new settled population there. The mitochondrial mother – the female who carried the original gene for this eye disease – was married in Quebec in 1669, and had five daughters and twenty-one great-granddaughters, all of whom carried the same deleterious DNA, and whose legacy is still borne by their male descendants to this day.

*

Intragenomic conflict can also arise via meiotic drive, which happens when some genes cheat by increasing their representation in the sex cells of the organism. All the genes in the bodies of animals and plants 'try'* to make their way into the reproductive germ-line

---

* Quick reminder: 'try' doesn't mean that genes have intentions or needs or wants but is a helpful heuristic for summarising a complex idea. As Richard Dawkins explained, 'we must not think of genes as conscious, purposeful agents. Blind natural selection, however, makes them behave rather as if they were purposeful, and it has been convenient, as a shorthand, to refer to genes in the language of purpose.'

– but the production of sex cells is a kind of genetic lottery. Sex cells are produced via convoluted process, called meiosis, which is a bit like shuffling a deck of cards before dealing it into two piles.* Each pile of cards corresponds to a single sex cell, and each card in a pile is a gene. Sex cells therefore carry only half the number of chromosomes (and genes) of the other cells in your body: the full complement of chromosomes is restored at fertilisation, when one sex cell meets another.

The creation of sex cells sparks the potential for conflict because only one sex cell will typically be fertilised. Any given gene therefore has a 50% chance of making it into the lucky sex cell that will eventually form a new individual. Any selfish genetic element that can stack the deck in its own favour will therefore outcompete co-operative genes who play by the rules. How might they achieve this? Some genes sneakily duplicate themselves so that they appear on every chromosome prior to segregation. In the analogy above, this is equivalent to a card making sure that it appears in both piles (therefore guaranteeing its appearance in the sex cell that ends up being fertilised). Other genes are silent assassins that identify and eliminate any sex cell where they aren't present. These murderous elements can result in reduced fertility by reducing sperm or egg count. One in seven human couples experiences difficulty in conceiving, and it seems reasonable to suggest that infertility might often be caused by selfish gene variants, like meiotic drivers (although identifying the actual genes in question continues to pose a formidable challenge).

One way of thinking about selfish gene variants is to imagine them as being like people who jump to the front of the queue, instead of joining the line at the back. Queue-jumpers benefit from

* Strictly, it is like *duplicating* a deck of cards before shuffling and dealing into four piles, but the maths works out the same and the description above is easier to follow.

their intolerable behaviour, but they inflict costs on the rest of the people waiting patiently in line. Just as a queue-jumper shows scant concern for the people queuing, selfish genetic elements don't care too much about the chaos they cause, so long as they get their own way. As a consequence, although selfish genetic elements can severely harm the host organism, their nefarious tactics mean that they can rapidly spread through a population if unchecked. One of the most effective solutions to a problem like this turns out to be cooperation.

To prevent the interests of the selfish few from corrupting the well-being of the collective, genes form a united front – a so-called 'parliament of genes' – where the combined efforts of the many prevent the corruption of the system by the selfish few. In the UK, political sparring occurs in the House of Commons, with debates being policed by the Speaker, who ensures that Members of Parliament respect the rules. There is no such authority figure in the genome; instead, fairness is enforced by majority rule. Going back to the queue-jumping analogy, genes work together much like the people waiting patiently in line might: they join forces to stop a 'queue-jumping' genetic renegade from unfairly getting ahead.

*

Other all-too-common maladies can also be better understood – and maybe even treated – by seeing them as genetic or cellular outlaws, pursuing their own agenda. Perhaps the most salient of these is cancer. The anxiety that the word cancer provokes is a relatively recent phenomenon. Since the dawn of the Agricultural Revolution about 12,000 years ago, living at high population densities has meant that the principal health concern for most humans came in the form of infectious disease. This concern is still highly relevant even now, as our experience with diseases like Ebola, influenza, and (of course) COVID-19 has shown. Nevertheless, thanks to spectacular advances in sanitation and modern medicine over the last hundred years or so, our species' susceptibility to infectious diseases has substantially reduced.

But as our ability to combat infectious disease has improved, the spectre of cancer has grown, with the most recent estimates predicting that half of us will be diagnosed with a cancer of some sort during our lifetimes. In 2020 in the US alone, it is estimated that more than 1.8 million cases of cancer will be diagnosed. That's more than 200 diagnoses per hour, or around three per minute.

Cancer is especially concerning due to its insidious, seemingly unavoidable nature: it lurks inside us, an enemy within. Cancer can occur when mutations arise in our body's cells, allowing them to evade the checks and controls that regulate normal cellular growth and activity. For a tumour to be cancerous, its cells must have several 'hallmark' features. The tumour needs to be able to generate its own growth signals, as well as the blood vessels that secure its nutrient supply. In doing so, the cancer cells can cream off the resources that would otherwise be used for normal cellular activity. Cancerous cells must also become immortal. Most cells in the body carry an internal 'clock' which keeps track of the number of cell divisions the cell has undergone. Once a cell reaches a certain limit, or age, it stops dividing. Cancer cells become ageless by switching off this internal clock. Powers of evasion are also paramount: cancerous cells ignore the signals in the extracellular environment that limit cell growth and disregard any instructions to self-destruct. Finally, cancer cells must be able to disperse to new tissues and set up a new tumour site there: this is what distinguishes benign from metastatic tumours.

Although cancer was originally thought to be a clonal disease (with cancerous tumours all carrying the same dangerous mutation) we now know that this is wildly improbable. A single mutation could not give rise to all of these abilities in one move. Instead, it is more likely that cancerous tumours actually consist of many different cell types, which act in unison to achieve these hallmarks. In other words, cancerous tumours are best understood as cooperative communities rather than selfish clones.

The first suggestions that tumours might actually comprise lots of different 'subclones' were voiced as early as the 1970s, but were met with disbelief and largely ignored for almost thirty years. This may have been a tragic oversight. It turns out, far from being the exception, most cancers are defined by heterogeneity – tumours are diverse communities where the different cell types help one another to thrive. For instance, the cells of subclone type-A might secrete a growth factor that can be used by the cells of subclone type-B. Type-B cells, in turn, might produce a molecule that inhibits responses to growth suppressors that type-A can also make use of.

We now know that the most aggressive and invasive forms of cancer are formed from these diverse, mutualistic communities. This can help to explain why drugs that seem to target tumours in one part of the body are ineffective in other areas, and how cancers evade therapy and re-emerge after seeming remission. Viewing cancer in these terms also offers suggestions for more effective treatments: we might be better able to target metastatic cancers by disrupting these partnerships, for example by targeting the cell types upon which other cells in the tumour critically depend rather than attacking the tumour as a whole.

This view of cancer also highlights a more general point: cooperation at one scale is often felt as competition at another. A community of cancer cells may cooperate with one another inside a multicellular body, but this cooperation exacts a heavy toll upon the host organism. There is a melancholy futility to this situation. Even a cancer that wins the battle eventually loses the war: most cancers are non-transmissible, meaning that they cannot escape the organism's body.* They may temporarily hijack the vessel for their own ends, but when the ship goes down, they perish with it.

*

The dynamics of this lose-lose scenario can help to explain what at first glance seems to be a contradiction: most multicellular organisms (including us) are extremely good at *not* getting cancer – but half of us will get cancer at some point in our lives. How can both of these statements simultaneously be true? The trillions of cells in each of our bodies must be regularly copied and replaced at a ferocious rate, around 100 million per minute. Every time a cell is copied, there is the potential for an error – a mutation – to creep in. Copying the genetic contents of a cell would be the equivalent of you copying out the entire contents of this book without making a mistake. Then doing it again. And again. When we think of it this way, it becomes remarkable to think that we don't develop a new cancer every day; cancer starts to seem less like bad luck and more of a genetic inevitability.

In reality, potentially cancerous mutations do show up from time to time during cell divisions but the genes in our body, which have had a long evolutionary history of meeting cellular outlaws, have evolved a series of check-and-control mechanisms to prevent these renegades from invading. Just like the different control points you face when trying to board an international flight – emptying your pockets, showing your passport – the cells in our bodies have genes that code for proteins with similar functions, checking that all appears as it should before allowing the cell to proceed on its division journey. For the most part, our bodies have the upper hand in the war against cancerous cells because our genes have seen it all

* Some cancers are infectious, although they are rare. A famous example is the Facial Tumour Disease that afflicts Tasmanian Devils.

before. In the same way that passing through airport security has become increasingly onerous in response to criminals finding novel ways to bring planes down ('Take off your shoes!', 'No liquids!'), cellular control mechanisms have upped the ante in response to their tangles with the genetic renegades of yore. Each potentially cancerous cell, on the other hand, arrives as a novel mutation within a body, with no genetic preparation or insight into how to best hijack the host's defence. Most of the time it is therefore, thankfully, trivial for the cellular security forces to capture and eliminate these invaders at source.

So why does cancer become more likely as people age? Here, we see the cold hand of natural selection at work again. The strength of selection wanes during the lifespan: so long as an organism successfully survives and reproduces, selection will not act too strongly against the properties of genes that allow later-life diseases to nose their way in and get a foothold. This insight cuts particularly close to the bone for our family: my mum was not yet forty years old when she was diagnosed with bowel cancer in 2004. She recovered, but was diagnosed with another, more dangerous tumour in her breast just under a decade later. This time the prospects look far bleaker. She will leave five children, two of whom have already had children of their own. By its own heartless logic, natural selection therefore declares her a winner. George C. Williams, the evolutionary biologist, was right – Mother Nature is a wicked old witch.

Part 2

THE FAMILY WAY

Humans don't come in ones. We like company – and, in fact, we *need* it. Feeling excluded from social interactions fires off pain signals in our brain – the same sorts of signals that also alert you if you burn your hand or break a bone – and loneliness is associated with a suite of insidious side effects, from disturbed sleep all the way through to reduced immune function and an increased risk of death. Togetherness is wired into us to an extent that makes it difficult to objectively assess the advantages to group living. But sociality comes at a price: living at high densities can increase susceptibility to disease, as well as increasing competition for resources. Animals that live in groups also have to relinquish some autonomy over their optimal schedule. The scenario of going out for dinner with a group of friends provides a loose analogy: you might have preferred to eat at a different time, or in a different restaurant, but for the sake of doing something together rather than alone, you need to compromise. For many species, like polar bears, the costs of group life are too intense and individuals tend to live mostly solitary lives.

So why bother joining up at all? Grouping offers a way to buffer individuals from the environmental challenges they face, sometimes conferring benefits that are sufficient to outweigh the costs. Emperor penguins form enormous huddles, thousands of birds strong, to stay warm in the bleak and bitingly cold Antarctic winters. In fact, the huddles are so effective that the innermost penguins end up overheating and have to nudge their way outwards; this constant rearrangement of the parts gives the colony the appearance of having an endless fluid motion when viewed from above. Group life can also offer protection from predators: the wildebeest that inhabit the open plains of East Africa live in massive herds because there is safety in numbers. This arises via an obvious dilution effect (if group size is N, then there is a 1/N chance that you will be the one who is picked off by a lion) and also a less obvious confusion effect, whereby predators attacking a herd can find it harder to lock in on one specific target, rendering the hunt less successful (or more, depending on whether you take the perspective of the predator or the prey!). Larger herds also mean that there are more pairs of eyes to scan the environment, increasing the chance that a lurking predator will be spotted before they strike – and reducing the time that any single individual has to spend on vigilance, rather than more profitable activities like foraging or mating.

Zooming in to observe cellular interactions hints that anti-predator benefits could have driven the ancient evolutionary transitions from free-living unicells to multicellular organisms. We know this because we can experimentally engineer the evolution of multicellularity from single-celled forebears in the laboratory, using just a glass tube, some single-celled algae and a bit of glorified pond water. Adding a predator that eats the single cells provides the impetus for the singletons to coalesce into a multicellular cluster – to form a group. Clusters are typically about eight cells strong, a size that seems to perfectly reflect the costs and benefits of grouping: small enough for each cell to maintain access to the nutrients in the liquid medium and large enough to avoid being munched by the predator.

For the most part, though, these kinds of groups are transient and fickle. Herd size waxes and wanes with the threat of predation, and even the engineered clusters of algal cells that evolve in the lab revert to solitary life when the predator is removed. Joining a herd to avoid being eaten by a lion is one thing, but being a member of a stable group with fixed membership is quite different.

Although social life seems like the most natural thing in the world from our anthropocentric point of view, we can start to appreciate that stable social groups are actually incredibly special – and something which result from a delicate evolutionary appraisal of the costs and benefits of living together. Many primates live in social groups, and humans are no exception. Nevertheless, we are unique among the great apes in that we also live in stable *family* groups, where mothers receive assistance from others in the production of young. The evolution of our family – fathers, siblings and grandparents – was the first critical step on our path towards becoming a hyper-cooperative species. In what follows, we will explore why humans developed families, starting with their most important members: mums and dads.

4
OF MUMS (AND DADS)

> I demur to your saying ... that animals are governed only by selfish motives. Look at the maternal instincts & still more at the social instincts. How unselfish is a Dog! To me it seems as clear that we have a conscience as that the lower animals have a social instinct: indeed I believe they are nearly the same – But these are mere trifles.
>
> Charles Darwin, 1860

I once heard the evolutionary psychologist, Robin Dunbar, quip that it is the first forty years of parenting that are the worst. Having enjoyed the exclusive company of my children for the last six months (due to pandemic-induced school closure), I get where he's coming from. But, even in normal times, we are a species that invests heavily in our young. Because it is so natural to us, it is not immediately obvious that parental care is a form of altruism and something that therefore warrants an evolutionary explanation. Protecting offspring or carrying them around or nourishing them with milk are all costly investments that have emerged, just like the complex eye, from an ancestral starting point of no care. But why bother looking after your offspring at all?

We can certainly imagine alternative worlds, where individuals maximise their reproductive success by having as many babies as they can, but investing very little in each of them. In fact, we do not need to rely on our imaginations as many parents *do* invest the bare minimum in their offspring. Parental care is uncommon in insects and other invertebrates, and many fish simply lay a batch of eggs that will be fertilised and then left to their own devices. Other species, like cuckoos and cowbirds, foist the tiresome parenting duties onto others by laying their eggs in foreign nests and scarpering, leaving the hosts to raise these bolshy intruders.

But at the other end of the spectrum, we see species like the black-lace spider that give everything for their young. This attentive mother carefully fashions an enclosed nest from a leaf and nurtures her brood for four weeks in the family home. When the spiderlings hatch, she provides a meal of unfertilised eggs for them to eat and, a few days later, allows the babies to devour her alive. The mother is not a passive victim of her cannibalistic offspring but instead actively encourages them to begin feasting on her. Offspring that receive this valuable meal are heavier when they leave the nest and experience better survival. By sacrificing herself, the mother provides her young with the best possible start in life, a spidery silver spoon.

Although the black-lace spider is an extreme case, investing in parental care of any sort means that the already expensive business of reproducing costs even more. Remember that resources are finite – every animal has a fixed amount that they can spend on reproduction and survival. It can be helpful to think of these resources as being funds in a savings account: spending now means you will have less to spend on other things, or in the future. Reproduction might come at the expense of survival, and investing in this breeding attempt might come at the expense of the next.* Parents are most likely to spend some of their metaphorical money in the bank when

* When a female nears the end of her lifespan, the trade-off between current and future reproduction is less acute because the female doesn't need to save any of this metaphorical money in the bank for the future. As a result, females commonly invest more in reproductive efforts that occur towards the end of the reproductive lifespan.

the costs involved are offset by the survival and reproductive benefits that accrue to offspring.

One of the most devoted parents that I know of is the yellow-billed hornbill, a bird I got to know while I was working in the Kalahari Desert. The hornbill has a slightly freakish appearance: mad-eyed and clumsy, with a distinctive, curved yellow beak. A few hornbills at our study site became quite tame, probably because there was a constant human presence, with dozens of scientists collecting data on various species around the reserve. Although I wasn't studying hornbills, I loved watching them, particularly in the breeding season. Before they breed, a male and female hornbill form a close and intense bond, with the male offering the female morsels of food and the occasional beakful of colourful flowers to boot. This courtship period is an important opportunity for the female to judge the male's likely prowess as a husband and parent, for she will be entirely dependent upon him if and when they do raise a brood together. Like all hornbills, the female nests inside the cavity of a tree trunk and subsequently seals herself inside this temporary prison to lay her clutch of eggs. She leaves just a tiny slit of a window, through which the male can pass items of food, sufficient to sustain her (and the chicks, when they hatch) during her almost forty-day confinement. The male hornbill therefore becomes the lifeline for the female and the young brood that hatch inside the tree, demonstrating the lengths parents will go to to ensure the survival of their young.

*

Assuming that evolution favours some degree of parental investment, who should actually pick up the tab for this expense? One of the most striking – and mysterious – patterns in nature is that, even in species where we cannot easily tell males and females apart by looking, we can often take a stab at identifying the sexes by looking at who is raising the offspring. With a few notable exceptions, where parental care occurs, it usually falls to females to do all or the bulk of the work. Males only stick around to help with offspring if their efforts confer a significant additional advantage to the young – sufficient to compensate the male for the lost reproductive opportunities he might otherwise pursue.

Of course, there are exceptions to this general rule. We've already seen that male hornbills are doting fathers and, in a minority of species, childcare duties fall entirely upon the male. This is quite common among fish where the eggs are fertilised externally, because a female can spawn her eggs on the male's territory and swim off, leaving him to take care of the kids. In birds, biparental care is the norm, though females commonly still invest more than males in offspring.

In human societies, mothers (on average) tend to invest more in offspring than fathers, an expense that begins while the infant is inside the mother's womb. Gestation is costly for all female mammals, but women seem to pay an especially high price. Compared to other primate species we give birth to large babies, though most of their brain growth is deferred to the postnatal period. As a result, newborn humans have the appearance of being born before they are ready to face the world. They have relatively poor cognitive and motor skills, are unable to see or hear as well as adults, and are completely dependent upon the adults who look after them for food, warmth and protection. Indeed, for a baby human to be born with the same cognitive and motor skills as a baby chimpanzee, it would have to be born an eye-watering nine months later.

Until fairly recently, the thinking was that women gave birth to these undercooked infants because larger babies, with bigger heads, would not fit through the mother's birth canal. But work published in the last three or four years tells a completely different story: human pregnancy is terminated because of energetic, rather than anatomic, constraints. Towards the end of pregnancy, a woman's resting metabolic rate becomes untenably high, more than twice that of a non-pregnant female. One recent study likened the metabolic demands of pregnancy as comparable to those experienced by runners during an ultra-marathon! Birth might therefore be triggered by the mother hitting this metabolic ceiling, rather than because an older baby with a larger head would not fit through the pelvis.

Of course, the energetic demands of parenting do not end at birth: children remain dependent upon caregivers for food and protection for many years. And, even after birth, mothers tend to invest more in their children than fathers do. It has become modish to assert that female-biased care is culturally entrenched, rather than biologically influenced, but a cursory look at our neighbours on the tree of life shows that we cannot lay *all* the blame for a mother's work never being done at the feet of the patriarchy. We are mammals – a group where, in more than 90% of species, parental duties fall entirely to the female. All female mammals gestate young internally and provision them with milk when they are born. There is less scope for males to be helpful during this time – and, also, more chance for them to abandon the mother, leaving her quite literally holding the baby. And the evidence is unequivocal: in every human society for which we have data, the presence of a mother has more important consequences for the long-term development and survival of a child than does the presence of a father.

That said, however, humans are among a small minority of mammals where males *do* invest in their offspring although the extent of this investment is quite variable. For instance, the Hadza and Datoga are two populations who live close to one another in northern Tanzania. Despite their geographical proximity, men in these two populations exhibit striking diversity in their approach to childcare: whereas Hadza men spend lots of their time holding, comforting and playing with their babies, Datoga men view childcare as 'women's work' and spend little to no time caring for or even interacting with their young children.

In humans, as in other species, paternal care will typically be traded off against pursuing additional mating opportunities: helping out with the kids means that males can put less energy into finding and mating with other females and siring offspring elsewhere. We might therefore expect to see specific hormonal profiles among men who are dedicated fathers compared to those who are less invested, or not currently caring for an infant. Testosterone is a promising candidate hormone for mediating these effects because it is commonly linked to mating efforts. We cannot manipulate testosterone in new fathers for obvious ethical reasons – but these experiments have been done in other species. In many (though not all) cases, injecting males with testosterone renders them more interested in mating and less interested in parenting. In humans, the next best thing to performing an experimental manipulation is to do a longitudinal study. This involves measuring the concentration of a candidate hormone at different points in time and determining whether variation in hormone levels map onto variation in a particular behaviour or trait of interest. Using such methods, researchers have found that men experience a reduction in circulating testosterone levels when their baby is born. What's more, fathers with lower basal testosterone levels tend to report spending more time caring for their infants and show increased activity in areas of the brain associated with the drive to nurture offspring. In societies where it is not customary for men to care for offspring (as is the case among the Datoga of Tanzania), no such change in testosterone among fathers is observed, which is also consistent with this hormone being involved in regulating paternal care.

*

But why is it that mothers so often pay the greater costs of investing in offspring? Naïvely, we might assert that females are forced to invest more because of simple biological differences: men cannot gestate young or lactate which would seem to force the female's hand. However, taking a broader look across the animal kingdom shows that this explanation is unconvincing. Female pregnancy is not a foregone conclusion: in seahorses, for example, males gestate the young. The bird equivalent of gestation – incubating a clutch of eggs – is a job that is normally done by the female but there are some species, like the ostrich, where these duties fall mostly or entirely to the male. As in all mammals, it is female humans that breastfeed infants, but in principle, it could just as easily be men. They have all the machinery and, if they receive the hormone prolactin, men can even produce milk. But they haven't evolved to do this. Saying that female mammals invest more because they gestate or lactate is a bit like saying that you ran out of money because you spent it all. It's a circular answer – what we really want to know is *why* females are the ones who make these costly investments (incubating eggs, gestating young, lactating) when, in principle, males could have evolved to pick up the tab for these parental care traits instead.

One of the main reasons that females tend to invest more than males in offspring is because it is often easier for a female to be sure that she is the mother of the offspring than it is for a male to be sure that he is the father. An unavoidable part of being a male is lower parental certainty, and investing in offspring that are not yours is likely to be a costly evolutionary mistake.

Males frequently go to great lengths to prevent rivals from copulating with their mates – sometimes even sacrificing their lives to this jealous pursuit. The orb-weaving spider is a grisly example. These spiders are a bit unusual: females have two receptacles for storing sperm, and males have two sperm-delivery devices, called palps. Ordinarily the female will only allow the male to insert one palp at a time, but sometimes a male manages to force a copulation with a juvenile female, during which he inserts both of his palps into the female's separate sperm-storage organs. If the male succeeds, something strange happens to him: his heart spontaneously stops beating and he dies in flagrante. This may be the ultimate mate-guarding tactic: because the male's copulatory organs are inflated, it is harder for the female (or any other male) to dislodge the dead male, meaning that his lifeless body acts as a very effective mating plug. In species where males aren't prepared to go to such great lengths to ensure that they sire the offspring, then the uncertainty over whether the offspring are definitely his acts as a powerful evolutionary disincentive to provide costly parental care for them.

*

There's another reason that females might often be left to do most of the work: they face intrinsic limits to reproduction that males don't seem to. This limit stems from the very definition of what it means to be a female or a male. Not all species have binary sexes but, where they do, we can tell the sexes apart by looking at the size of the gametes (sex cells) individuals produce: females produce few, large eggs and males produce lots of tiny sperm. This is why we know that it is the male seahorse that becomes pregnant, rather than a more circular method of decreeing that the pregnant animal must be the female. Whereas eggs contain nutrients to help the fertilised zygote to grow, sperm often supply nothing more than a snippet of genetic material. Sperm cells are sexual parasites – letting the egg provide all of the nutrients and gaining a free ticket to the gene pool in doing so.

Egg cells tend to be costlier than sperm cells to produce, and females therefore tend to produce fewer sex cells than males. This means that males and females face different limits to their reproductive success. As an analogy, consider what might happen if all the available egg and sperm cells were at a ball and had to find an opposite sex partner to dance with. The eggs, by virtue of their scarcity, are almost certain to find a dance partner if they want one; whereas many of the sperm cells would be left watching on the sidelines. This analogy can be applied to thinking about limits on reproductive success in the real world. Whereas most egg cells can find a sperm partner, the majority of sperm cells never get to fertilise an egg. Male reproductive success is predominantly limited by how many eggs they can fertilise – by access to females. Females, on the other hand, are limited not by access to males but by the stark ecological currencies of time and resources. Even in humans, the most reproductively successful men far outstrip the achievements of their female counterparts, with the most prolific fathers reportedly siring more than 1,000 children compared to the record of sixty-nine children for a single mother.

These reproductive constraints can have profound consequences for the strategies males and females use to secure their genetic legacy. As a general rule, male reproductive strategies should be more likely to emphasise quantity over quality, whereas females should (again, on average) be fussier about who they mate with, only investing their precious reproductive resources in high-quality males. But the patterns we observe in nature are more nuanced than this. The extent to which males are expected to prioritise mating with multiple females (polygyny) versus sticking with the female partner they have (monogamy) depends critically on how easy it is to find new female partners. One factor that affects this is the sex ratio: the abundance of females, relative to males.

When females are plentiful, we should see increased competition among males to mate with as many females as possible (and to provide little other input into offspring). Conversely, when females are in short supply, then it can be more profitable for a male to stick with the female partner he has, if he finds one. Thinking back to the analogy of eggs and sperm at the dance may be helpful to grasp the logic underlying these predictions: at a ball with ten sperms and a hundred eggs, a sperm who waltzed briefly with each egg partner would have danced many times by the end of the evening. At this ball, polygyny is a winning strategy. But what happens if we flip those ratios around, so that there are ten eggs and a hundred sperms at the ball? Here, a 'love them and leave them' strategy is unlikely to pay off. If you happen to be one of the lucky sperms that does have a partner to dance with, your best bet is to stick with her (and to fend off the advances of the unpartnered sperms), rather than trying your luck at finding another.

In evolutionary terms, several things can happen once males are committed to mate-guarding a female. A male that stays with his female partner, rather than leaving to seek additional mating opportunities elsewhere, might then benefit by directing some of his energy into the care of offspring. This shows how paternal care can evolve as a by-product of males sticking around to defend their female partners from other males (rather than the need to help raise costly offspring). Wide-ranging phylogenetic analyses, which have constructed the evolutionary history of mating systems and paternal care across more than 2,500 mammal species, suggest that this order of events is likely to be a general pattern: monogamy came first – as a way to defend females from would-be competitors – and fathers followed.

In some shorebirds, like jacanas, males are so abundant relative to females that females undergo a complete sex-role reversal, becoming larger, more aggressive and dominant to males. Female jacanas don't bother with any of the parental care, simply laying their eggs in the male's nest and leaving him to take care of the details. In other species, like burying beetles, males calibrate their investment in offspring according to the strength of competition from other males. These remarkable insects get their name because they bury dead mice and other small vertebrates: they are the grave-diggers of the insect world. This is not done to preserve the dead creatures' dignity, but instead to provide a starter home for the burying beetle babies. Burying beetles are one of the only insect species that care for their offspring by provisioning them with food. Newly hatched beetles can feed directly upon the carcass they are born in

– but it is more efficient if the parents feed the young, which they do by regurgitating bellyfuls of digested meat for the offspring to eat. The drive to protect paternity seems to be critical for explaining paternal care in burying beetles – males stay with the female and their offspring when they face competition from other males, but are quick to desert the female and the brood otherwise.

What about humans, though? An abundance of males might turn burying beetles into better fathers but that doesn't mean that the same goes for us. Moreover, although we can experimentally manipulate beetle sex ratios in the laboratory, we obviously can't conduct these sorts of experiments on humans to test how increasing numbers of men affects their tendency to invest in paternal care.

We can, however, make use of the fact that humans have often conducted these sorts of 'experiments' over the course of our history. In Australia, adult sex ratios were artificially manipulated in the late 1800s due to the unequal numbers of men and women who were deported to this nascent colony as convicts: in some locales, there were as many as sixteen males for every adult female. In those areas where it was raining men, women were more likely to be married and provided for (meaning that women were less likely to participate in the labour force). One interpretation of these results is that when men were plentiful, they seemed to be investing more, not less, in relationships with women. In fact, modern-day gender roles and attitudes in these parts of Australia can be traced back to these historical sex-biases: in those areas that previously had a surfeit of men, women are less likely to work and people of both sexes espouse more conservative attitudes regarding what constitutes 'men's' and 'women's' work.

A more recent study of the Makushi group, who live in Guyana, found something similar. As a consequence of men moving to remote areas for work in forestry or mining, and women preferring to live in urban areas, some communities find themselves with a relative shortage of men. In the urban areas, where they were in shorter supply, men reported being more interested in casual sex and less interested in committed relationships. In remoter areas, men described themselves as less promiscuous and more committed to a partner.

In general, many studies indicate that men are more willing to settle down and marry – and that conjugal stability is greater – when men are relatively abundant compared to women. However, as we shall see, even where parents do cooperate to raise offspring, there is likely to be conflict over the amount that each parent invests in the young, with the general rule being that each parent would like the other to do more of the work. How are these parental squabbles resolved?

5
WORKERS AND SHIRKERS

> Nobody will ever win the battle of the sexes. There is too much fraternising with the enemy.
>
> Anonymous

Zebra finches are hardy little birds that eke out their existence in the harsh Australian outback. Like many bird species, zebra finches form a strong pair-bond and, when the babies arrive, both parents *seem* to work tirelessly delivering food to the nest full of screaming chicks. But there is something puzzling going on: chicks that are raised by two doting parents end up being fed less (and are therefore lighter) than those who are raised by their mother alone. How can we explain this?

Parental care is a scenario that is ripe for conflict – even if both parties help to raise young, each has the temptation to invest slightly less, perhaps making two visits for every three made by the partner. Experimental work in zebra finches has shown that females slack off when they have a reliable male partner, leaving him to do more of the hard work of rearing the offspring. This female tactic results in the rather paradoxical outcome described above, where the offspring do worse when they have two parents compared to when they have just one. One can immediately envision how such strategies could result in a race to the bottom: if each parent is trying to undercut the other, then the poor chicks might eventually end up with no feeding visits at all. So, how are these kinds of conflicts avoided? Without being able to talk about it, how do parents negotiate the workload with one another and cooperate to feed the young?

Rather than each parent trying to undercut the other, theoreticians predict that if one parent's efforts drop off a little, the other should actually compensate for this shortfall by increasing their own workload. But – here's the important part – they should not compensate fully. To illustrate why, imagine two parent types, called 'Shirker' and 'Worker' (also known as 'Mum' and 'Dad' in our house). If Worker picks up all of the shortfall when Shirker takes a break, then there is really no evolutionary incentive for Shirker to help at all. Shirker might as well leave Worker to care for the young, and seek additional mating opportunities elsewhere. However, if Worker raises their game a little, but not enough to fully cover the shortfall in care, then Shirker is more incentivised to stay and help, because the offspring will suffer otherwise. Amazingly, many studies have shown that bird parents conform to the predictions of these evolutionary models: if you temporarily remove one of the parents, or make it harder for them to feed at the nest by adding weights to their tail feathers, the partner will work a bit harder, but will not fully compensate for the shortfall in care. Without being able to talk about it, birds therefore seem to negotiate conflict over parental investment.

Disagreements over who should invest in the offspring will be most pronounced where the male and female only expect to stay together for the current breeding attempt. The reason for this is easy to grasp: if a male only expects to have one breeding attempt with a female, then he has no real interest in her future reproductive potential. In other words, a male would prefer that the female spends all of her resources on *this* breeding attempt where she will raise *his* offspring – he doesn't want her to save anything for future breeding attempts where the offspring may be sired by another male. The conflict will be attenuated where the male and female expect to stay together for a longer period because the male has more of a vested fitness interest in his partner's future reproductive potential under these conditions.

We can measure how these disagreements over who ought to do the most work play out in the real world by documenting the ways in which males and females manipulate one another into providing more care. Female burying beetles are cunning manipulators of their male partners. These beetles work hard to rear their young, by regurgitating bellyfuls of digested meat for the babies to eat. When offspring are very demanding, the hard-working female is rendered temporarily infertile. (The same sometimes happens in breastfeeding humans, due to the release of prolactin, and is known as lactational amenorrhea.) Nevertheless, the male beetle might still try to solicit mating attempts with the harassed mother. Mating is costly to females and to prevent the male from pestering her when she is temporarily infertile, the female secretes a chemical which acts as a potent anti-aphrodisiac, putting the male off the idea of copulation. As a beneficial side effect, the male becomes more focussed on parenting the brood. A win-win for the female!

*

In many human societies, monogamy is the norm and you might therefore predict that sexual conflict would be kept to a minimum. However, the road signs here tell us to proceed with caution. First, we must be careful about inferring the ancestral human condition based on what we observe in contemporary groups: current practices may not be a reliable guide to where we came from. It is also unclear which human society our estimates of human mating markets ought to be based upon. Marriage and breeding practices vary quite broadly across cultures: although pair-bonded monogamous unions are the norm, polygyny (where men have multiple wives) is also common and a minority of societies are polyandrous (where women have more than one husband). Finally, when it comes to aligning male and female interests over the whole lifetime, there is quite an important distinction between monandry (females mating with just one male for their entire life) and serial monogamy. Under the latter scenario, although a male and female are pair-bonded for each breeding attempt, we might not see their fitness interests perfectly coinciding because an individual might acquire a new partner following divorce or the death of a spouse.

We can gain further insights into the ancestral human condition by examining anatomical features: differences in the body plans of males and females often reflect the underlying mating system. For example, gorillas live in polygynous societies, where a single dominant male defends access to a harem of fertile females. In such species, males tend to be larger than females and may also be more heavily weaponised, which helps males to fight off competition. If mating success depends on being chosen by a female partner rather than winning aggressive contests, then males can evolve to be brilliantly coloured and decorated and to perform impressive displays to woo female admirers (like male peacocks).

Another important clue to the mating system can be gleaned from examining testis size. To put it bluntly, where females are promiscuous (in that they mate with multiple males) then males tend to grow bigger balls relative to their body size. Larger testes can produce more sperm, allowing their male owners to more effectively compete in the mating arena. Examining testes size in the other great-ape species highlights the point. In gorillas, all females in a group mate with the dominant male, the silverback. As we would expect based on the low rates of sperm competition, gorillas have tiny testes relative to their body size. Contrast this with chimpanzees, where females are more promiscuous and mate with many males during an oestrous cycle. Whereas the gorilla has a walnut, a chimpanzee has a chicken's egg, these larger testes capable of producing almost 200 times as much sperm as the gorilla's.

We can use these anatomical features to make some inferences about the ancestral mating system in our own species – which should help us predict the scope for conflict between males and females. Men are typically larger and stronger than women, but are not weaponised like red deer or as dimorphic as gorilla silverbacks. This suggests that although there has been some selection on body size and strength in men, our male ancestors were not like gorillas and did not defend harems of women from other men. Instead, the disparity in body size between men and women is more in keeping with a largely monogamous species, where males and females are usually much more similar to one another, in both size and appearance. Although contemporary humans didn't evolve from a highly promiscuous ancestor, our mating markets were probably a bit more colourful than lifelong monogamy. In terms of (relative) size, human testes lie between those of the gorilla and the chimp, but are far closer to the gorilla end of the spectrum. From this we can infer that sperm competition was less important among ancestral humans than it is in chimps, but that human females likely mated with more than one male during their lifetime. One plausible conclusion seems to be that ancestral humans were monogamous, but serially so. Under serial monogamy, male and female interests are not wholly aligned over the entire lifespan, and we should therefore expect to observe some disagreements when it comes to parental investment.

*

Sometimes, the conflict between males and females can pervade the most intimate of relationships: that which exists between the mother and her unborn child. Pregnancy is viewed as a special time of bonding between the mother and child, and a time before the real battles of parenting start – the unborn foetus cannot cry all night or throw temper tantrums or do any of the other things that kids do to make their parents' lives difficult. But viewing pregnancy as a harmonious time is to see it through a fairly rosy lens. The stretch marks many of us carry bear witness to the costs of harbouring a rapidly growing foetus. These superficial scars also tell the tale of a much deeper conflict, an internal battle that is being waged between the genome of the foetus and the genome of its mother.

Like thousands of other women, when I was pregnant with my first child I was called to the hospital for a routine test to determine my blood sugar levels. This was one of the least enjoyable tests I underwent during pregnancy: I was required to fast for twelve hours and then to drink a lukewarm syrupy liquid, which I choked down in the largest gulps I could manage. After about an hour, a nurse pricked my finger to see how my blood sugar levels had responded to this glucose challenge. One of the other women in the waiting room vomited soon after she had forced down the last mouthful of goo, meaning that she had to be sent home and go through the entire procedure again another day.

Though it is unpleasant, this test is important for detecting gestational diabetes, a serious complication of pregnancy where a woman becomes unable to regulate her blood sugar levels, leading to the excessive growth of the baby with potentially life-threatening consequences. Gestational diabetes is a disease that can stem from conflict between the maternal and paternal genes inside the growing foetus. Each foetal cell contains genes that come from the mother and genes that come from the father. Some of these genes come with a marker that tells the gene which parent they came from – these are called imprinted genes. These imprinted genes can regulate gene expression, determining how much the effects of a particular gene are turned up or down.

Gene expression can explain why, despite all the cells in your body having the same genome, they perform wildly different functions depending on where in the body they happen to live. Cells in your mouth have to produce amylase (an enzyme contained in saliva which allows you to digest complex carbohydrates into simpler sugars) but I don't need – or want – the cells making up the skin of my arm to start drooling. Not every gene is equally expressed in every cell or organ of the body; instead, genes can be switched on or off, or turned up or down, depending on the part of the body the cell finds itself in.

Now here's the curious thing: not all the genes in your body agree about whether and how much they ought to be expressed. In the foetus, the genes that come from the mother and the genes that come from the father might squabble over how much they want to squeeze the mother to yield nutrients to the foetus. Although both sets of genes agree that the foetus should receive *some* resources from the mother, they disagree over how much. A useful to way to conceptualise this conflict between mothers and their unborn babies (or between the maternally- and paternally-derived genes, if you prefer) is as a tug of war. You may have engaged in negotiations with a young child over how much television they ought to watch. A bit of screen time affords you some peace and quiet or a moment to get some jobs done; but the child might wish to watch television for three hours, rather than a few minutes. The zone of conflict is the range where you disagree – and it is within this zone that the child is expected to employ tactics to persuade you into letting them watch more than you want them to. A similar zone of conflict exists between maternally- and paternally-derived genes in the foetus, with the latter demanding more investment than the maternal counterparts want to supply.

To see why the genes within the foetus might disagree, it helps to take the gene's perspective. The maternally-derived genes care about this baby's survival, but they also have shares in the future offspring produced by the mother. It is not in their interest to bleed the mother so dry that she is unable to produce any subsequent offspring. Paternally-derived genes, on the other hand, care more about the foetus than they do about the mother because they don't necessarily have a shared interest in the future offspring from that female. Thus, paternally-derived genes are selected to turn up the heat on the mother, selectively expressing in regions that are related to resource transfer in the placenta. These genes code for hormones that increase nutrient concentration in the maternal blood and can even modify regions of the brain that manipulate the mother's behaviour, driving her to provide more care for the offspring once it is born.

The front line in this conflict is the placenta: the nutritional interface between mother and offspring. Placental morphology varies hugely across mammals – mostly with respect to how invasive the placental cells are. Some species, like horses, have 'epitheliochorial' placentas, which politely respect the boundaries between their own tissues and the mother's. By contrast, humans and other primates have invasive 'hemochorial' placentas, which snake through the wall of the uterus and even burrow into the maternal blood vessels. Placenta cells are foetal rather than maternal in origin, and they therefore act in the interests of the foetus, not the mother. Human placental cells bathe directly in maternal blood, meaning that the mother loses control over the nutrient supply the foetus receives. In humans and other apes, the placenta is able to decide how many nutrients it will take, rather than the mother retaining control over how much she will give. Allowing the foetus the upper hand in this way may have evolved as part of an evolutionary compromise in species that tend to have fewer, high-quality offspring throughout their lives. Unfettered access to maternal blood supply makes it easier for a foetus to receive the nutrients it needs to grow to a healthy size inside the mother (and might be especially linked with the need to grow a large brain, as in humans), but comes with the attendant downside, from the maternal perspective, of loss of control.

The fact that mothers surrender control leaves them vulnerable to the actions of paternally-derived genes that act to increase nutrient supply to the unborn child. These genes typically work by coding for hormones – differential gene expression can therefore alter the hormone levels circulating in the mother's blood. For example, human placental lactogen inhibits the action of insulin. This elevates glucose concentration in maternal blood and also reduces the mother's ability to use this sugar herself, meaning that more of it can be siphoned off by the foetus. Other hormones increase the mother's blood pressure, which speeds the delivery of the nutrient-rich blood to the foetus. If unregulated, the actions of these hormones can have pathological and even fatal consequences for mothers. Higher blood sugar levels increase the risk of gestational diabetes and can also permit the unchecked growth of the foetus, creating potential problems during childbirth. Hormones that increase maternal blood pressure leave her vulnerable to pre-eclampsia – one of the most serious complications a mother can face during pregnancy.

*

A lesser-known risk of pregnancy is that it increases maternal susceptibility to cancer. In particular, deciduoma malignum – a cancer that arises when placental tissues continue to proliferate inside the uterus – occurs in women who believe they are pregnant (but aren't) or who have recently aborted. It can quickly be fatal if left untreated. Trophoblasts, the placental cells that invade the maternal uterus, share an alarming number of features with metastatic cancer cells, including the tendency to rapidly proliferate, to invade tissues, and to ignore instructions to undergo programmed cell death. Foetally-produced human-chorionic gonadotropin – the hormone that delivers the second blue line on a pee-on-a-stick pregnancy test – can also be produced by tumour cells and is present in up to 30% of human cancers. Their invasive nature means that trophoblasts sometimes hijack the wrong maternal tissues, leading to ectopic pregnancies, where a fertilised egg implants and grows outside of the uterus. Ectopic pregnancies most often occur in the fallopian tubes, but occasionally also in the mother's ovaries, abdomen, bowel or even, very rarely, in the site of the scar from a previous Caesarean section. Trophoblasts share so many similarities with cancerous cells that ectopic pregnancies can be treated with methotrexate, which is a chemotherapy drug commonly used to treat cancer.

Compared to other mammals, humans are especially susceptible to cancer. In part, this stems from our longer lifespans (cancer becomes more common as we age) and from our modern lifestyles, which introduce risk factors in the form of high-calorie diets and exposure to carcinogens, such as air pollution and tobacco smoke. Another, less obvious risk factor may be our placentas. Invasive placentas, like ours, are the ancestral condition in mammals, but have been secondarily lost in some species, including horses and other hoofed mammals. Downgrading the foetus' access rights might be a by-product of selection acting on females to protect themselves from metastatic (invasive) cancers. Because placental cells operate in such a similar way, their privileges are revoked as a by-product. As predicted by this hypothesis, recent evidence shows that metastatic cancers are less common in species with less invasive placentas. The upshot is that humans are caught in an untenable situation: we need invasive placentas to maximise nutrient delivery to our resource-hungry offspring – but opening the door to placental cells involves weakening our intrinsic defences against any kind of invasive cells, including cancerous ones.

*

After they are born, babies often continue to demand more resources than parents want to give. The most common complaint for parents of newborn children is sleep deprivation; this seems to be a cultural universal rather than Western epiphenomenon of different parenting practices. Newborn children frequently wake during the night and cry, prompting the mother (or father) to provide them with milk. You might not think there is anything strange about this at all: it could just be the child letting the parent know that it needs feeding and the parent dutifully responding – something which serves the interests of both parties equally well. However, a closer inspection reveals several curious features of infant night-waking that suggest there is more going on.

Frequent breastfeeding can render a woman temporarily infertile – a phenomenon called lactational amenorrhea (as I mentioned earlier). A woman who is breastfeeding regularly is therefore less likely to conceive, which allows the nursing infant a longer period before it has to compete for maternal resources with a younger sibling.* A child who wakes frequently during the night can, on average, expect to enjoy a longer period of uninterrupted care from the mother, which raises the possibility that human infants use frequent night-waking and breastfeeding to compete with future siblings that aren't born yet. If this hypothesis is correct, then we might expect the genes associated with night-waking to be paternally-derived, in the same way that those associated with transferring resources to the foetus at the expense of the mother also are. Why? Because the paternally-derived genes in the foetus care less about the future offspring the mother might produce than the maternally-derived genes do.

This is a pretty outlandish hypothesis but there is some evidence to support these predictions. Prader–Willi syndrome and Angelman

* There are several accounts from older anthropological studies of small-scale societies describing how newborn infants could be killed if the mother had given birth to twins, or if an older sibling was still dependent on the mother's milk. Extreme resource scarcity might not worry many people who live in modern industrialised societies but, for people living more precarious lifestyles, nutritional constraints might mean that mothers are unable to nurse a new infant at the same time as the older sibling.

syndrome are two genetic disorders that result from a deletion of part of the genome on chromosome 15. For Prader–Willi syndrome, the genes that are deleted are paternally-derived, meaning that children with Prader–Willi syndrome only have the maternal genes being expressed at that part of the genome. Curiously, infants with Prader–Willi are weak breastfeeders, have a feeble cry and sleep a lot. Angelman syndrome seems, in many ways, to be the opposite kind of disorder. Here, it is the maternally-derived cluster of genes that is deleted, with the consequence that the paternal genes are expressed. Infants with Angelman syndrome are renowned for being poor sleepers and waking frequently at night to breastfeed. Both of these syndromes are associated with a host of other serious difficulties and problems for the children in question, which brings us back to the tug-of-war analogy. Cases like this, where imprinted genes from one parent are deleted, are analogous to one team suddenly letting go of the rope – a contest that was previously more or less balanced can go seriously awry.

There is another, more exotic manner by which foetuses can manipulate maternal investment. This is by fluidly exchanging cells with their mother while they are in the womb. About half of us – maybe more – contain cells from another human being inside our body. As a woman, there is a reasonable chance that I contain cells with male Y chromosomes, cells that carry the exact same genetic stamp as those inside the bodies of my two sons. They, too, might have some of my cells swimming around inside them somewhere. As well as the cells from my children, I probably also have some of my mother's cells – and perhaps even my maternal grandmother's as well. We are chimeras.

What are those cells from my two boys up to inside me? Research on micro-chimerism is in its infancy and there is still much to discover. Some scientists have suggested that the cells from previous offspring might mediate sibling rivalry, by increasing investment in the current child and decreasing any investments in subsequent offspring. Preliminary data is highly suggestive: foetal cells commonly migrate to maternal breast tissue, where they might increase maternal milk supply to the infant. Research in rodents has shown that foetal cells also make their way to the mother's brain, prompting the growth of new neurons and possibly therefore affecting the ways females behave and interact with their offspring. Other work finds that the phenomenon of secondary miscarriage – whereby a previously fertile woman becomes unable to conceive or gestate a subsequent pregnancy – is associated with the presence of foetal cells from older brothers inside the mother's body, as is pre-eclampsia, suggesting that the foetal cells might play a role in extracting more resources from the mother and warding off competition from future siblings.

I am hesitant to write much about this phenomenon here as it is still early days, scientifically speaking. But it is worth mentioning simply because it is so mind-bendingly bonkers. It wouldn't surprise me if, in the next five or ten years, you might be able to read a whole book on micro-chimerism and the ways that the many cells from different individuals interact inside a single body.

Together, these patterns underline the point that cooperation typically goes hand in hand with conflict. The bonds between husband and wife – and especially between mother and child – are often viewed as devoted and sacrosanct, but even these closest of unions are plagued with subterfuge and struggles, as each party's genes try to tip the balance in their own favour. Conflict is therefore a fact of life, not just between warring parties, but also between individuals whose interests are mostly aligned.

Until now, my emphasis has been on the myriad ways that offspring can make their parents' lives difficult. But to end the story here would be misleading indeed. In the next chapter, we're going to meet the species where offspring become important – and helpful – members of the family. These are the cooperative breeders – and we are among them.

6
WELCOME TO THE FAMILY

One touch of nature makes the whole world kin.
William Shakespeare, *Troilus and Cressida*, 1602

It is natural to conceive of parents helping their offspring, but it is perhaps less well appreciated that, in some species, offspring also return the favour. This familial arrangement is called cooperative breeding, and, although it isn't common, it is a trick that has been discovered by evolution many times across the animal kingdom, including in ants, bees, wasps, termites, crustaceans, mammals, birds and fish. Humans are also part of this elite and rather diverse club, where mothers can expect to receive help from their older children in the business of rearing younger ones. And we are unusual because we are the only ape where this happens.

For those of us living in modern, industrialised societies, it might come as a surprise to discover that we are cooperative breeders, as we typically have relatively small families, and often stop breeding before the older children can become helpers to younger ones. I grew up in an unusually large family – the eldest of six children – so I had ample opportunity to show my worth as a helper, though I didn't always excel in this role. My mum often recounts the story of a time she left six-year-old me in charge of my younger brother while she did a household chore. I was a total bookworm as a child and had my nose buried deep in the pages of my story, oblivious to the goings-on around me. By the time my mum returned a few minutes later, my brother had managed to find the dog's food dish – and was busy rubbing its contents into the living-room carpet. In response to her admonishment ('Nichola! I thought I told you to watch your brother!'), I peered over the top of the pages and replied, 'I did for a while, but my book was more interesting.'

*

A more expansive view of humans around the globe suggests that unhelpful offspring are the exception rather than the norm. By the time children are about seven years old, until they reach about fourteen, they frequently help to take care of their younger brothers and sisters, both by babysitting them and by foraging for foodstuffs that can be brought back to camp and shared with the family. In fact, human children grow up surrounded by other family members, including siblings, aunts, uncles, cousins and grandparents, all of whom might be involved in helping to raise them. This situation is quite different to the other great apes, where offspring are reared almost exclusively by their mother and tend not to form special bonds with other relatives.

If you were to overlay the distribution of cooperatively breeding species onto a map of the world, you would notice that they are clustered in some of the harshest regions: the meerkats and mole-rats of the African deserts, the white-browed babblers and apostlebirds of Australia's outback, and the groove-billed anis and cotton-top tamarins of Central and South America. For early humans,* things were similar: for most of our existence, we have inhabited some of the planet's most difficult environments. The founding members of the *Homo*

---

* By which I include all members of the genus *Homo*, and not just our own species, *sapiens*.

genus lived somewhere between 1.5 million and 2.8 million years ago (around 6 p.m. on 31 December in our calendar-year analogy) in the East African Rift Valley. Many of the most important fossils for reconstructing the timeline of human evolution have come from this region, which sits on a fault line in Earth's crust where two tectonic plates are being pulled apart. Over the last 10 million years, this area has morphed from a flat, tropical forest into a jagged landscape, punctuated with dizzying mountains and vast basins, the latter forming the Great Lakes of East Africa. As the tectonic plates heaved apart, the East African Plateau was formed, bringing changes in weather patterns. An area which would once have been covered with lush forest became increasingly barren and dry. Against the backdrop of this progressive aridification, there were also dramatic climate shifts, linked to the emergence and rapid* disappearance of the Great Lakes in the basins that had formed from the plates' slow separation.

These early environments would have posed a number of existential threats. For one thing, food was probably quite hard to come by. In dry regions with sporadic rainfall, individuals must actively search for their meals. The human diet is thought to have consisted of fruits, plants (including potato-like tubers that had to be dug out of the ground) and meat (that had to be hunted and killed).† This is a risky foraging niche (buried tubers are hard to find and big game fights back) and the techniques take time to perfect. Working as a team and being able to learn the skills from others would have been crucial for survival.

Our ancestral habitat was also dangerous, being occupied not only by megaherbivores (food!) but also by the full complement of ferocious predators. Ancestral *Homo* may have looked much like you

---

* Rapid in geological time. The lakes are thought to have appeared and disappeared in a cyclical fashion every 200,000–400,000 years.

† The other apes – including chimpanzees – are mostly vegetarian. Chimpanzees do hunt and kill other animals, but meat comprises just 5% of the calories they ingest, compared with over 50% for modern hunter-gatherers.

or I, but to the sabre-toothed cats and gigantic lions* that prowled the savannah, I suspect that our human relatives probably looked a lot like lunch. By this point, humans were committed to a terrestrial mode of life and the increasingly exposed landscapes of the African grasslands likely made it difficult to flee to cover when confronted with danger.

We had one thing going for us though: we had each other.

As we saw with the herds of wildebeest, safety in numbers can emerge simply from dilution and confusion effects. It seems likely, however, that humans took things one step further: we learned to fight back against predators and drive them away. Observations of other primate species, like chimpanzees and baboons, support this hypothesis. Primate species that are more exposed to predation risk, because they live on the ground or inhabit barren habitats, tend to live in larger groups with more males. More importantly, there is also evidence that, in these groups, males counter-attack predators and even sometimes subdue them – for example, chimpanzees regularly attack predators and have even been observed to corner and kill leopards. It is likely that our ancestors also did the same.

Our evolutionary climb down from the trees made us more effective in combat with predators. Because we walked on two legs, our hands were free to hold things – like weapons. And because our arms were no longer required for swinging through the branches, the musculature could become specialised for throwing rocks and other projectiles from afar. In addition to helping us to fend off predation threats, these adaptations allowed early humans to become effective scavengers, driving predators away from meat and appropriating the spoils for ourselves.

With such skills at our disposal, it was a small jump for humans to become predators ourselves. The development of these abilities

* Recent fossil evidence suggests that Pleistocene lion species were significantly larger than modern lions.

coincided with the great migration of *Homo sapiens* out of Africa, and the subsequent extraordinarily rapid colonisation of the planet which, in our 'time on Earth as calendar year' analogy, took around eleven minutes. As we moved around the globe a trail of megafaunal extinctions followed in our wake, documenting our dramatic rise from the bottom of the food chain to the top.

There is a simple conclusion that we can draw from this whirlwind tour of early human evolution: we needed to cooperate to survive. This helps to explain why there is almost no evidence in the fossil record of other apes living alongside humans in the East African Rift Valley. Instead, our great-ape cousins inhabit less seasonal and more plentiful environments where extreme cooperation is not a prerequisite for survival. One can't help but wonder, if we were less like humans and more like chimpanzees, whether we might now exist only as yet another data point in the *Homo* fossil record, as relics of a species that tried and failed.

*

Given the advantages of teaming up, it might seem surprising that a rather small percentage of species breed cooperatively. For instance, there are no known amphibians or reptiles that breed cooperatively. For insects, spiders, mammals and fish, the numbers hover at or below 1% of species, and it is around 8% for birds. Cooperation is difficult to achieve and, for it to flourish, the conditions have to be conducive. Helpers are most likely to emerge in species where females are monogamous and can produce multiple offspring with each breeding attempt. Monogamy stacks the deck in favour of helping because helpers will be assisting in the production of full, rather than half, siblings. Likewise, there is more scope for helpers to increase their mother's reproductive output when females produce multiple offspring with each attempt, rather than singletons.

Humans are an outlier in this respect, as we tend to give birth to one child at a time, and our reproductive schedule seems to prioritise quality over quantity. We've already seen that gestating even one child places a heavy metabolic burden upon human females, so mutations that increased the likelihood of giving birth to twins (like some other cooperatively breeding primates do) would involve each foetus receiving less, or would demand an unfeasibly high investment from pregnant mothers. We took a different route to make use of the help on offer: rather than increasing our litter size, we simply pumped our offspring out in quicker succession. In contemporary forager societies, mothers that receive assistance in the production of young are able to initiate a subsequent pregnancy once infants are weaned, meaning that they can reproduce approximately twice as quickly as our closest ape cousins: whereas chimp offspring are separated by periods of around six years, hunter-gatherer females can give birth to (and raise) a child around once every three years. Cooperative breeding therefore allowed human mothers to resolve the quality/quantity trade-off, by giving birth to extraordinarily high-quality offspring, and in relatively high numbers.

Our status as cooperative breeders has important implications for our understanding of human society and parenting norms. Our social way of life means that, for most of our time on Earth, mothers have been embedded in vast social networks and children have been raised by multiple caregivers, including fathers, older siblings, aunts and uncles, and grandparents. Many contemporary human societies still live like this, though these large extended families have (to some extent) been replaced by more formal institutions, like schools and day care, in many industrialised societies. Formal institutions that provide childcare are a logical extension of our cooperative breeding natures, and the fact that we are cooperative breeders is likely why they exist in the first place.

Unlike the other great-ape species, where infants remain with their mother at all times, human children frequently spend more time with other caregivers than with the mother – and need not form a unique bond with one caregiver but might instead be looked after by several individuals, who might even be children themselves. This perspective stands in stark contrast to the Western ideal of the nuclear family, where a parental unit is expected to raise young with little assistance from the extended family group, and where the tendency to have fewer babies closer together means that any older siblings are typically still dependent, rather than able to act as helpers.

This is an area where insufficient attention to our evolutionary history can have pernicious consequences. The Western model is frequently used as the benchmark for best practice, informing policy decisions about what parental care *should* look like, and how we might damage our children if we don't adhere to it. Attachment theory is one of the more prominent ideas in this space. Under this world view, a child's healthy development depends on its ability to become securely attached to a principal caregiver – usually the mother. Following this logic, it is argued that mothers who are insensitive or unresponsive to their offspring (or who send their children to daycare) can change the course of their child's development, leading to what is known as an insecure attachment style in the child, with the potential for far-reaching negative consequences.

The implicit conclusion is that mothers are responsible for the production of socially responsive, well-adjusted children who are able to form functional relationships in adulthood and go on to become productive members of society. When this goes wrong it is also, therefore, mothers who are to blame. The problem with this theory is that the nuclear family, with its emphasis on one parent caring and the other working, is very unusual, both cross-culturally and when considered in broad historical and evolutionary perspective. The idea that mothers are the exclusive and irreplaceable caregivers for children is a modern, Western cultural ideology, not a scientifically backed imperative.

The hard data supports this view. From 1991 to 2007, the National Institute of Child Health and Development (in the US) followed more than 1,000 children, starting when they were just a month old and ending when the children were in Grade 9 at school. Some of these children went to day care, while others were raised exclusively by the mother. The report's main conclusion was unequivocal: children who went to day care and those who stayed at home with a parent did not differ in their development. Similarly, a more recent study conducted from 2003–6 with more than 1,400 French children found that those who attended a good nursery typically had fewer emotional and conduct problems than those who were cared for by mothers alone, perhaps because the children attending nursery had more opportunities to interact with peers, and other adults. These studies don't downplay the importance of the mother–child bond, or say that mothers have no role to play in ensuring the healthy emotional development of their offspring. Rather, they encourage us to consider childcare in a broader perspective, incorporating multiple caregivers and relationships, a perspective that better reflects our long evolutionary history as a profoundly social and cooperative species.

7
YEARS OF BABBLING

> In my simplicity, I remember wondering why every gentleman did not become an ornithologist.
>
> Charles Darwin, *Autobiographies*, 1876–81

In an effort to better understand how cooperative breeding works, I spent my early twenties chasing birds around the desert. Not just any birds, though – I was interested in one species in particular, pied babblers. These portly black and white birds live in intensely social groups in the Kalahari. I was fortunate to have been one of the first people to study their behaviour in detail. The focus of my research was to understand the conflicts that inevitably arise in a society that is built on cooperation.

By far the most common social arrangement among birds is the pair-bond, where a male and a female team up to raise a brood of nestlings. For many species, this marriage of convenience lasts only for as long as is needed, the union dissolving when the chicks fledge the nest. Babblers are different in that they are permanently and totally committed to a social way of life: they do everything – forage, sleep, play, and work – *together*. Group size is variable: sometimes babblers live as a tight-knit trio but larger mobs of up to fourteen birds are also common. Within each group the right to reproduce is monopolised by a single male and female. Everyone else is relegated to helper status, their prime responsibilities being to feed, protect and generally care for the dominant pair's offspring.

My research took place at the Kuruman River Reserve, a site that was established by my PhD supervisor, Professor Tim Clutton-Brock, and dedicated to long-term field studies of animal behaviour. The reserve sits on a remote piece of farmland in a part of the desert known, euphemistically, as the 'Green' Kalahari. In reality, the Kuruman River is a brown, sandy bed, that flows just once every twenty years or so. The Kalahari sand runs through several shades: dirty-brown near the riverbed, fading out to straw-yellow and eventually turning to the deepest ochre as you move further into the dunes. It isn't a verdant place, despite the name, but it has its own beauty.

The Kalahari is a rather harsh habitat to eke out an existence, and cooperation seemed to be the norm rather than the exception. The pied babblers I studied shared their desert home with several other social species, including bat-eared foxes (an endearing grey canid, with comically large ears), social spiders (identified by their wispy cotton-wool nests), meerkats, sparrow weavers, ants (of many varieties), and sociable weavers. The latter species builds giant super-nests, made up of many smaller cavities, all connected to one another like an avian block of flats. New arrivals tack their nest onto the existing super-structure until, eventually, gravity wins and the whole thing comes crashing down. Despite their many differences, these social creatures had a shared interest in something: the weather. In the desert, life depends on rain.

In the summer months, the first suggestion of rain would come with the arrival of a north-west wind, bringing with it the distinctive smell of wet earth and rain that had fallen far away. As dark clouds gathered, we would eventually be treated to the most spectacular storms, with enormous crashes of thunder and the sky coming alive with forks of lightning. In the bush, skeletons of trees that had been struck bore witness to the intensity of these events. Although they say lightning never strikes twice, the house we lived in was hit on several occasions (often causing the phone lines to blow out as well as frying any computer or device that had accidentally been left plugged in).

After a storm, the desert would burst into life. Red sand was newly carpeted with yellow flowers; sticky, stinging sour grass brushed at our legs; and the animals – released from survival mode – could turn their efforts to breeding. Rain in the desert is good news for the creatures that live there. Rain means food – and this food comes in plagues. First came the caterpillars, which would emerge in their thousands a few days after a big rain. The caterpillars always seemed to be on an upwards mission. In the field, they would climb trees, though we would constantly be plucking them from our legs, our hair and our dinner as they inched their way up in the world. Logically enough, some weeks later we would usually receive a deluge of moths. These gathered in great fluttering crowds outside the house, clamouring at the light on the veranda. Woe betided anyone who opened the door of the communal farmhouse without first ensuring the inside lights were turned off: a moth cloud would rush into the room, choking everyone with the powder that puffed from their rattling wings.

But the desert rain is a fickle friend, and these occasional storms punctuated long periods of drought. Life in the desert is a precarious existence. When times are good, they're really good; but when they're bad, they're awful. A cooperative way of life can buffer individuals from the extremes of this harsh environment. Rather than going it alone, it can be helpful to be part of a team that works together, profiting in the good times and riding it out together in the bad.

*

Although we were less reliant on rainfall, feast and famine were also familiar themes for the human inhabitants of our field site. In addition to the babbler researchers, the field site was also home to a dozen or so volunteers and a handful of scientists who were running the much larger project: a long-term study of meerkat behaviour. For us, food availability was dependent not upon rain but on infrequent shopping trips. The reserve was a bumpy three-hour drive on unsealed roads from the nearest town, and we typically only ventured in once a month, buying enough food to see the twenty-odd inhabitants through for the coming weeks. Anything we didn't get, we wouldn't have that month. Fruit and fresh vegetables were always the first items to run out, or to perish in the heat. Chocolate and beer never lasted long. Towards the end of every month, we would inevitably find ourselves eating meals produced mainly from tinned or dried goods, including the infamous and widely detested 'cheesy spaff' – a loose take on lasagne that had presumably been invented by a desperate volunteer of yore, when tasked with providing the evening meal from the meagre remains of that month's food.

But the babblers faced another peril that we humans were thankfully spared: as well as the difficulty in finding their own food, they had to avoid becoming someone else's lunch. The desert is an open sandy expanse, and there are few places to hide from predators. Babblers are particularly exposed because they forage on the ground, using their beaks to swish away piles of sand and flush out the small invertebrates hiding just below the surface. This 'head down, bum up', technique leaves them vulnerable to predators, especially to aerial threats from birds of prey, including giant eagle owls, pale chanting goshawks (or PCGs as we called them) and martial eagles. PCGs stood out because of their bright orange legs and matching beaks. They commonly perched on the wooden fence posts that ran for hundreds of kilometres alongside desolate roads, waiting for their next unsuspecting victim to arrive. The martial eagle was an even more formidable adversary: a large bird with a vast two-metre wingspan – though I only ever saw them as black pinpricks through my binoculars as they soared above the baked land below. On one occasion at our study site, the radio collar of a missing dominant female meerkat was traced to the nest of a martial eagle. As well as the radio collar, the nest contained the bones of a duiker, a small antelope that inhabited our field site, demonstrating the ability of these apex predators to dispatch sizeable prey.

*

The babblers' fear of predators initially presented a huge impediment to being able to observe their behaviour in the wild because the birds were also highly suspicious of us. We could scarcely get within a hundred metres before they would spot us, uttering high-pitched alarm calls and fleeing far away. We realised our best chance of having the babblers accept us was to locate their nests, and to sit motionless some distance away while the birds made feeding visits. But habituating a wild animal to your presence is a slow and extremely boring process. We would sit for hours in the sand, intermittently inching closer, aiming to push the boundaries without scaring the birds. If you ever find yourself in this situation, choosing a good spot to sit is vital: many a time I accidentally stationed myself too close to an ant nest and spent the next three hours being attacked by irascible insects as I disrupted their activities. Moving was not an option: every time I stood up, the babblers would be freshly alarmed by my presence and would flee the area for several minutes, undoing all of my hard work in an instant. My mission was to blend as unobtrusively as possible into the background, to the extent that the babblers stopped noticing that I was there.

After several days of quiet sitting – and once I was able to sit close enough to attempt it – I would try to throw mealworms to the babblers. The trick here was to land a mealworm right under a babbler's beak as they foraged. If my aim was true, and the babbler ate the mealworm (two things which did not co-occur as frequently as I would have liked), I would also whistle. The aim was to train the babblers to associate a tasty food item with the undulating whistle we used and with our unthreatening presence. Over a period of months, we gained the birds' trust, meaning that we could observe their natural behaviours close-up. The whistle training was also a success: rather than spending hours searching for babbler groups in their vast territories, we could whistle and, in return for a juicy mealworm each, the whole group would fly to us! We also trained the babblers to jump on a portable balance in return for a sip of water, which they sipped in a most comical manner – as if it were a fine wine – from the lids of our water bottles. Weighing the birds meant we could obtain longitudinal measures of condition and figure out who was doing well and who was faring poorly.

The aim of our research on babblers was to explore their cooperative way of life and to determine who did what in these tight-knit family groups. But there was an immediate problem: to the human eye, all pied babblers look alike. To keep tabs on what individuals were doing, we also needed to be able to identify them. We did this in the usual way: by colour-ringing every bird in the population with a unique four-ring combination. As well as giving the birds a unique ID, meaning we could keep track of them, the rings meant we could also give each babbler a nickname which made discussion among ourselves easier. For instance, one of the longest-standing matriarchs in the population was a female with white and red rings on her right leg, and white and metal on her left. Her ID was therefore WR|WM, and we called her Wonder Woman.

At the start of the project we had to individually trap the adult birds one by one to ring them, a procedure that inevitably damaged our relationship with the babblers and renewed their distrust in us, sometimes irreversibly. To avoid this, once we had ringed all of the adult birds in the population, we kept tabs on where the groups' nests were and, rather than individually trapping and ringing the adult birds, we ringed the babies while they were still in the nest. Ringing the nestlings meant that we could avoid 'dehabituating' the adult birds, but we did so at great risk to our own personal safety. Accessing babbler nests usually meant tiptoeing on the final rung of a ten-metre ladder, while reaching into the branches of a hostile acacia tree to extract the nestlings from their thorny home. Often, we could not see into the nest, instead having to reach in and blindly fish the nestlings out of the interior. The omnipresent worry of falling from the ladder was thus sometimes accompanied by an equal concern that we would place our hands upon a venomous snake rather than a brood of nestlings (particularly if we arrived at the nest when no adults were nearby making feeding visits and we were uncertain whether the chicks were still alive).

*

Once we were able to observe the babblers without them perceiving us as a threat, we rapidly realised the extent to which the babblers' efforts were directed towards protecting themselves (and the vulnerable chicks) from danger. I confess that initially I had the impression that the babblers were overly cautious about predators. Often, they would squeak in alarm at seemingly inexplicable threats – a plane flying overhead, or a bird of prey so far in the distance I could barely see it, even through my binoculars. One day, however, I witnessed first-hand how quickly a member of the group could be lost to a predator. I was collecting data at a group called Sox – a breeding pair that had successfully fledged a pair of chicks, despite having no helpers in the group. The adults were run off their feet by the insatiable, bleating chicks, and they foraged feverishly, trying to match the supply to the demand. Busy with our respective tasks, neither I nor the babblers realised that a yellow mongoose had managed to sneak up on us. All of a sudden, it burst out from under a thorny bush and snatched one of the chicks. It was all over in seconds.

In larger groups, babblers take it in turns to watch out for predators by performing sentinel duty. Sentinels take up a high position

– a tree branch or a fence post – and scan the environment for danger. Meerkats also do something similar and it became a rite of passage for the volunteers at the project to obtain a photo of themselves acting as a human sentinel post, with a meerkat perched on top of their head. While on guard, the babbler sentinel gives a reassuring watchman's song, a low-pitched 'chuck' that tells the foraging group members that everything is safe. If the sentinel spots a threat, they let out a warning call – with different calls denoting different kinds of predator – and the foraging group can quickly dash to safety.

Perhaps because of their cooperative sentinel system, the babblers always spotted predators far more quickly than me – and some of those predators were also things that I needed to avoid. As a consequence, I always felt far safer walking around the desert when I was with the birds than when I was alone. I recall one particular occasion – a sultry afternoon that I had spent trudging round a babbler territory whistling to no avail. There had been a large rainfall a few days before – the babblers were always less responsive to our whistle and the prospect of a mealworm when their usual food supplies were plentiful. It was hot and – unusually for the desert – humid: the damp sand seemed to steam under the glare of the sun. We called it snakey weather as the moist heat seemed to encourage them to emerge from their hiding places, bringing them out to hunt during the day. My eyeline was fixed on the trees, hoping to spot the familiar little blobs of white and black, when I heard a slight rustle in the bush at my feet. The cobra was no more than a foot from me, its forebody already raised off the ground with its hood erect: a sign that the snake feels threatened and is ready to strike. The cape cobra is a snake you don't want to provoke: it has a dangerous combination of traits being both nervous and highly venomous. I froze and backed away slowly, and the snake dropped its guard and slinked off.

The babblers were far braver than me. If they saw a snake, they would frequently harass the creature, gathering round and emitting loud, jarring caws in an attempt to drive the predator from their territory. Several other dangerous adversaries, including mongooses and owls, were also mobbed in this way. As we saw earlier with humans, babblers succeed by working together.

*

Babblers need to be careful while foraging, but by far the most dangerous time to be a babbler is in the nest. Babbler nests are usually built high up in thorny trees but, aside from being difficult to access, their open-cup structure offers little tangible protection. Frequently, the entire brood of nestlings was lost to predators. This peril was particularly acute overnight and sometimes the dominant female, who always incubated the nest overnight, would also perish. Due to this grave risk, the babblers have several cooperative tricks up their sleeve to reduce nestling mortality, studies of which formed the basis of my PhD. Groups with insufficient helpers are less able to spend time protecting the nest and repelling predators from it. As a consequence, the smaller groups encourage their chicks to fledge the nest very early, when the chicks are still underdeveloped and unable to fly. Despite being rather feeble, these chicks are safer out of the nest than they are while they are in it. Larger groups, who are better able to protect the nestlings, allow the young a valuable extra few days in the nest before encouraging them to leap from it.

One of the more elaborate ways that babblers collaborate to protect the nestlings is by coordinating their feeding visits at the nest, going in pairs rather than alone. Though initially puzzling, we eventually realised this was a clever way of concealing the nest from predators. Here's why. Babbler nestlings spend the majority of their time hiding silently in the nest. However, every time an adult arrives at the nest with a food item, three hungry chicks reach up out of the nest, begging loudly for the morsel on offer. Each feeding visit is therefore a potential unwanted advertisement to predators: 'THERE IS A NEST OF CHICKS RIGHT HERE!' By coordinating their feeding visits, babblers have discovered an ingenious solution to maintain feeding activity at the nest, without increasing the number of times the chicks inadvertently broadcast their location to predators. An adult who finds a food item will fly to a tree branch and wait – for several minutes if needed – until another individual joins them. Then, they fly off to the nest together. This cooperation allows the birds to deliver two or more food items to the nest for the price of one bout of loud (and potentially dangerous) begging.

Things get even more interesting when the offspring leave the nest. Babbler offspring are completely dependent upon the adults for food for several weeks after they fledge. While in the nest, the babies spend most of their time hunkered down, bursting into life when an adult arrives with a food item. Once they fledge, however, everything changes. Now the chicks run along the ground after the adults, screaming loudly and incessantly while the adults search for food. One can't help but sympathise with the frazzled birds when, apparently out of sheer frustration, they occasionally jump on top of a bleating chick and give it a severe peck on the head, earning a brief respite in the process. Work led by a PhD student showed that babbler chicks can be quite devious, and exploit their vulnerability to predators to coerce the adult caregivers into providing more food. When the chicks are satiated, they happily sit in a tree, relatively safe from predators. But when the adults aren't supplying food quickly enough, the chicks move to the ground, placing themselves in danger and forcing the adults to feed them at a higher rate than they otherwise would. This kind of 'blackmail' strategy was first predicted by the famous Israeli biologist, Amotz Zahavi (who, incidentally, worked on another species of babbler in the Negev Desert). The chicks' willingness to put themselves in mortal danger by flying to the ground where they are most vulnerable to predators acts as a reliable signal of their hunger and convinces the adults to provide food at a higher rate.

*

One helping behaviour that I became especially interested in is teaching. Teaching is something we think of as being a quintessentially human activity. And though it might not be apparent at first, teaching is a form of cooperation: an instructor goes out of their way to help a pupil learn a vital skill or gain important knowledge. Though it doesn't always occur in formal settings, like classrooms, the process of helping others learn, whether by giving them opportunities or by active instruction and feedback, occurs in every society and culture on Earth. Teaching is the way that we learn skills like reading and writing. But teaching also plays a vital role in the emergence of cumulative culture – the tendency for societies and technologies to increase in complexity over time. This ratcheting effect can only be achieved if each generation can stand on the shoulders of the last, rather than having to continually reinvent technologies and rediscover important facts and know-how from scratch. In fact, teaching is so fundamental to the evolution of human societies as we know them that, before 2006, it was thought to be unique to humans – one of those behaviours that sets us apart from other species on the planet.

We now know that isn't the case.

In fact, examples of teaching in the animal kingdom are plentiful – but they don't always crop up where we expect them. When we think about species that might do things similarly to us, we tend to turn to the great apes. And yet, though there's good evidence that young chimps are adept at *learning* new skills and tricks from their mothers (for example, how to use a stone tool to crack a hard nut), there's no concrete evidence to suggest any active teaching on the part of the mother.

To understand why not, we have to think more generally about what teaching is – and what it is for. Teaching is a special form of helping behaviour. As with all kinds of help, there is a cost involved with providing the tuition. Teaching is most likely to have been favoured by evolution where the benefits of investing outweigh the costs. In the case of chimps, there are probably no convincing examples of teaching because the numbers just don't stack up. Young chimps are prolific social learners. They learn by watching and doing and don't seem to need active instruction. The skill of cracking a nut with a rock, or using a stick to fish termites from a nest (both of which have been proposed as examples of teaching in chimps), can be acquired independently through trial and error, with little danger to the juvenile if he occasionally gets it wrong. The benefits that could arise from teaching are also questionable: nuts and termites are relatively niche items in a chimp's diet. For the most part, chimps subsist on fruit and leaves, for which special handling skills are not needed. It follows that the cost to mothers of providing tuition is unlikely to be reimbursed in the hard currency of increased survival or reproductive success of her offspring.

Viewing teaching through this lens helps to explain why the most convincing examples in nature occur not in our closest living relatives, but in species where the benefit-to-cost ratio is highest. Unlike chimpanzees, who acquire the vast majority of their daily calorie intake from easy-to-find foods such as fruit and leaves, early humans occupied a more complex foraging niche, relying on foods they had to either extract (e.g. buried tubers, or nuts inside shells) or hunt. These more complex foraging techniques take time and skill to learn – and cannot easily be acquired through observation alone. The combination of foraging skills being difficult to learn and necessary for survival in humans may be the point of difference between us and the other great apes, explaining why we are prolific teachers while our ape cousins are not.

In fact, the first species to knock humans off our teaching pedestal wasn't a primate, a mammal, or even a bird. It was an ant.

*

In 2006, Professor Nigel Franks and colleagues stunned the scientific community with the revelation that *Temnothorax* ants teach one another the quickest routes to a food source or a new nest location. Ants can carry one another – and it would be far quicker for a knowledgeable ant to simply carry naïve individuals to the desired destination. But ants who are carried cannot learn the way for themselves (because they are carried facing backwards). To memorise the route, an ant must walk there itself, making frequent small loops to learn the various landmarks along the way. The leader acts as a tutor, going slowly, waiting for the pupil to make its many excursions, until they reach the destination. Once she has learned the route, the pupil can become a teacher herself.

There are now other convincing examples of teaching in several species, but – conspicuously – none in non-human primates. The second example, hot on the heels of the ants, came from meerkats. Meerkats are highly cooperative mongooses that, like the babblers, live in extended family groups in the Kalahari Desert. As in many cooperatively breeding species, reproduction is largely monopolised by a dominant breeding pair, who receive assistance from subordinate helpers. One form of help they provide is tuition. Meerkat helpers teach pups how to handle dangerous prey items, like scorpions, by providing pups with opportunities to practise on live, but disabled prey. When handled effectively, a scorpion is a high-quality lunch item. In the wrong paws, however, a scorpion is a potentially deadly adversary, its sting packed with toxins that are potent enough to kill even a human. Opportunities for pups to practise handling scorpions are scarce and fraught with danger. To learn what to do, the pups must be taught.

And that's exactly what happens. When the pups are very young, they are fed with freshly killed food items. No skills required. However, as the pups get older, the helpers start to up the ante. A maimed gecko. A scorpion with its tail nipped off. Eventually, helpers provision young with live, intact prey. These lessons give pups the chance to practise handling mobile and potentially dangerous prey in a relatively safe manner. Meerkat lessons are costly to deliver because provisioning pups with live prey takes longer – and the prey is more likely to escape. If a helper's only goal was to get as much food as possible down the pup's throat, then provisioning with dead food items would be more efficient. But the benefits of allowing pups to hone their hunting skills outweigh the costs.

*

Teaching is only worthwhile if the pupil is truly naïve: there is no point investing in teaching if the pupil already knows what he's doing. But how do teachers infer the ability or knowledge of pupils? In humans, teaching often relies on language and our ability to look under the hood, to 'know what the pupil knows'. This ability is part of a suite of cognitive mechanisms that come under the collective banner of theory of mind. For a long time, the assumption that animal teachers would also need these cognitive superpowers hindered our search for examples of teaching in non-humans. None of the examples I've mentioned so far occurs in species that have such abilities or are regarded as being particularly brainy. In fact, as Alex Thornton and Katie McAuliffe cleverly showed, meerkat helpers use a very simple heuristic to tailor the lessons pups receive. Rather than keeping track of pups' abilities, meerkats simply adjust the difficulty of the lessons to the pup's age – which they infer because older pups' begging calls sound different to younger pups' calls. To demonstrate this, Thornton and McAuliffe tricked the meerkats by playing pup begging calls through a speaker. Proving that they are in fact quite easily fooled, the meerkat helpers often ran over and dropped food items in front of the begging speaker. Meerkats who heard the calls of older pups tended to bring more intact scorpions, whereas those who were played the calls of young pups tended to bring dead, or de-stinged specimens. So, meerkats *do* teach but they clearly do it in a different way to humans. More generally, when ecology favours it, we find that many species arrive at the same behavioural destination as us, despite taking a completely different cognitive route to get there.

Other examples underline the importance of considering the utility of teaching when searching for convincing cases. Cheetahs probably teach their young in much the same way as meerkats do. This makes sense given the stark requirement for an adult cheetah to be able to hunt successfully and the high risk involved in bringing down prey: a well-placed kick from a wildebeest or gazelle could seriously injure or even kill. Cheetahs are typically solitary and – unlike lions, who live in groups – cannot scavenge food caught by other individuals. If they want to eat, they have to hunt. In addition, cheetahs are not particularly big or strong and cannot rely on brute strength to bring prey down. Rather, cheetahs kill by delivering a bite to the jugular and hanging on for dear life while the victim asphyxiates, sometimes for as long as ten or twenty minutes. Learning to kill with such precision requires practice and, like meerkats, cheetah mothers give their cubs opportunities to learn, by provisioning them with prey that has been injured but not quite finished off. If you have a pet cat, you might also have been the unwilling pupil in a similar kind of lesson – the tendency to release a not-quite-dead mouse in the kitchen is also likely the expression of this kind of teaching behaviour.

*

While I was working on babblers, teaching was one of the first behaviours I observed, though at the time I didn't realise it. I had been observing feeding visits at a nest of one of our newly habituated groups, collecting data on who was feeding the chicks and how much. When the nestlings were about eleven days old, I started to notice the adults giving a guttural rolling 'purr' call. A few days later,

I revisited the nest and observed that the nestlings now responded vigorously as soon as the adults emitted this purr call, even if the adult was on a lower branch of the tree and not yet visible to the nestlings. This was the first indication I had that the nestlings might have learned to associate the purr call with the arrival of a feeding adult. If that was the case, then this could be an example of teaching – the first example in any bird species. What's more, the adults would be conditioning the nestlings in a Pavlovian manner to associate the call with food delivery!

But, of course, there were other scenarios I had to rule out. The most obvious alternative explanation was that the chicks' response to the call was not learned, but simply emerged at a fixed point in development. If the chicks weren't learning anything, then this wouldn't be a case of teaching. I decided it would be interesting to test this. To do so, I ran a series of experiments where *I* became the teacher. My hypothesis was that if the chicks had to learn to respond to the calls, then providing additional lessons should help them to learn faster. If I was wrong, and responses to these calls emerged at a fixed point in development, then no amount of additional instruction from me would matter.

To provide these extra lessons, I first made recordings of the adults giving their purr calls. I then played these calls back to nestlings via a little speaker that I sellotaped to the end of a broom (all very high-tech!), giving me the means to broadcast at roughly the same height as the nest. I identified six nests where I would provide the extra lessons – playing the purr calls to chicks during every feeding visit. I started my lessons slightly earlier than the babblers did, playing the calls from days nine to eleven. The procedure for the six control nests was identical, except that I played the calls at random rather than during feeding visits. The crucial test would be what chicks did in response to purr-call playbacks on day eleven. I expected the chicks in the 'extra lessons' group to respond, by begging and reaching out of the nest. For the chicks in the control group, who had not learned to associate these calls with food delivery, I expected no response to my playbacks. That's exactly what I found – suggesting that babbler nestlings learn to associate purr calls with food delivery – and adults are responsible for teaching them.

The babbler example is a bit of a strange one, though. For the meerkats and the cheetahs, the value of the lessons is obvious: learning to handle prey is a vital life skill. However, it is far less obvious how babbler nestlings might benefit from learning to associate an arbitrary call with food delivery. I subsequently discovered that this training was an elaborate ruse, used by the adult babblers to trick the youngsters. We saw earlier that the adults have to encourage the chicks to jump from the nest – and that smaller groups exhort the chicks to fledge a few days earlier than larger groups do. The problem is that the chicks aren't that keen on leaving the nest. This is where the purr calls come in – the ones that the nestlings have been taught to associate with an incoming food delivery. To encourage the chicks to leave the nest, to fledge, the adults perch a little way away from the nest and start giving purr calls, despite having no food to offer. The chicks don't know this though. Hearing the adults calling them for dinner encourages the chicks to climb to the edge of the nest and, eventually, to take the plunge (quite literally – they cannot fly and often fall from a great height when they fledge). Once the chicks have fledged the adults continue to use these 'white lies' to bait the chicks into following them, for example to the evening roost tree, or away from predators. Teaching therefore turns out to be another of the babblers' rich adaptations to dealing with a predator-rich environment, and protecting their young.

*

Studies like these are painstaking, involving many hours of patient trudging and observations of what is often quite mundane behaviour. But if we want to learn more about the other species on this planet, we have to put in the time to get to know them. If an alien observer had been watching me for the last few hours as part of a study of human behaviour, it might conclude that we are a sedentary species that mostly ignores its offspring and stares intently at a glowing screen. To really understand the richness of human behaviour, the alien would have to stick around, to observe me at different times and in different contexts.* The same is true when we want to discover more about the social lives of other animals; a failure to do so risks presenting an artificially simplistic view.

The several established and long-term studies of animal behaviour mean that we now know a lot about the cooperative ways of babblers and meerkats and baboons (and a whole host of other interesting species). This is valuable because it allows us to appreciate how much we have in common with other social creatures, and because it highlights the tremendous diversity – in form and complexity – that cooperation can take. As we've seen with the babblers, cooperation frequently entails *doing* different things: looking out for predators, protecting the nest and providing offspring with food and tuition. But occasionally cooperation involves *becoming* different, by making permanent and irreversible alterations to oneself to become a better helper. We'll see how in the next chapter.

* On second thoughts, maybe the alien wouldn't be so far off the mark.

8
IMMORTALS

> Some pirates achieved immortality by great deeds of cruelty or derring-do. Some achieved immortality by amassing great wealth. But the captain had long ago decided that he would, on the whole, prefer to achieve immortality by not dying.
>
> Terry Pratchett, *The Colour of Magic*, 1983

Family life can be hectic, with many jobs that need to be done. A worker ant's daily schedule might include defending the nest from attackers, tending an underground fungus farm, or rescuing a sister from the jaws of a predator. A meerkat helper might perform sentinel duty to watch for predators, babysit newly born pups at the burrow, or even produce milk to feed these babies. Social species are usually territorial, and helping to defend the group's territory from intruding neighbours is also part and parcel of group life.

When there is a lot to do, it can be more efficient if individuals specialise in particular activities. Specialisation can lead to economies of scale – this is the reason why fast-food restaurants are, well, fast. You don't see staff members in McDonald's running around trying to do everything; instead, larger tasks are divided up into smaller, specialised roles. One person mans the till, another puts the order together, someone fries the chips, someone else makes the burgers, while different people tidy the restaurant and sweep the floor. Division of labour is the principle upon which factory assembly lines are built. It is more efficient for one person to become totally specialised on a small task, than for everyone in the factory to try their hand at everything.

In social species, the workload is often shared among group members in similar ways. In every group, there are some helpers that are more diligent and others that are lazier and, in a few species, workers do different jobs at different stages of life. Among honeybees, the safe jobs inside the hive are performed by the younger members of the colony, while the more dangerous task of foraging in the outside world is left to the older workers. Losing an old bee who is near the end of her lifespan anyway is less consequential, from the colony perspective, than losing one of the younger bees who would otherwise be available to contribute for a longer time.

Although age-based roles are quite common in cooperative societies, it is less typical to observe workers becoming totally specialised for one job for their entire life. We can understand why by taking a closer look at the exceptions to the rule. It is among the most highly social (or 'eusocial') insects – ants, termites and clonal aphids – that we see a striking point of departure. In some ant species, workers develop into a morph whose sole job it is to detonate themselves. These suicide-bombers are just one kind of highly specialised worker present within the colonies of the aptly named exploding ant *Colobopsis explodens*. If the colony comes under attack, the exploders spray the target with a toxic, yellow goo from their ruptured abdomens. The nozzle-headed termite is similar: the workers' heads are filled with nasty liquid that can be squirted at enemies. In the honeypot ant, workers develop into living food stores, their rear ends becoming so swollen with honey that they are rendered immobile. A gentle stroke of their abdomen encourages these living vessels to secrete the stored honey to provide food for their sisters. More recent work has also shown that some aphids develop as 'sticking plaster' morphs, whose job is to explode and thereby seal any cracks that appear in the colony wall – just like the cells in our body form scabs to seal a wound.

We've already seen that ant colonies can be likened to multicellular organisms. Division of labour among workers is another way in which ants seem to behave like cells in a body. These are jobs for life. Just like a cell in your heart cannot change its mind and decide to become a liver cell instead, once an ant has committed to becoming an exploding morph or a living honeypot, there is no going back. In fact, we only reliably see extreme and irreversible commitments of this sort in the most supremely social insects, because these societies operate as super-organisms rather than colonies of individuals.

There is, however, one kind of activity in a social species that is frequently split among individuals in the group: the job of breeding. Like many cooperatively breeding species, the babblers and meerkats of the Kalahari tend to have just one reproductive pair in each group. But helpers only delay rather than entirely forego reproduction. Subordinate babblers and meerkats have all of the physiological apparatus they need to be able to breed but are prevented from doing so, either by the actions of the dominant breeders or because they lack access to unrelated breeding partners. Helpers can therefore be thought of as hopeful reproductives, individuals that keep their options open and wait for the day that they might themselves attain a coveted breeding position.

But at the more extreme end of the spectrum, helpers and breeders are assigned their status at birth and subsequently follow completely different developmental trajectories. When this happens, those that are destined to inherit the crown become highly specialised for their important job as chief reproducers. The queen of the social termite species *Macrotermes bellicosus* is a grotesque distended blob, more than thirty times the size of the dainty workers who serve her. A honeybee queen grows to be twice as large as the workers – her size allowing her to spew out over 1,500 eggs per day. Her gargantuan size also traps her within the hive, as she is unable to fly. As the colony prepares to swarm, the workers place the queen on an exercise and dietary regime (by feeding her less and also nudging and shoving her so that she keeps walking around the nest) in order to slim her down to the point that she too can fly the nest.

Unlike the hopeful reproductives in babbler and meerkat groups, the workers in termite and ant societies are permanently sterile: they simply cannot breed, even if favourable conditions arise. This is really quite remarkable: we would feel indignant, perhaps even outraged, if we were ordained to live as modern-day eunuchs, our only purpose being to advance the lot of the chosen few. Yet some species *do* adopt this strange way of life, and we are, in fact, among them. We even have a special name for the sterile morphs in our societies: we call them grandmothers.

*

Since having children of my own, I have become aware of slipping a few places down the hierarchy of people that my own parents want to hang out with. Like a smash hit from a new pop star, the grandchildren have entered the charts at number one. But as rewarding as grandparenting may be, their existence is an evolutionary conundrum: why does human female reproduction cease so long before death?

Hardly any other species on Earth has such a prolonged post-reproductive lifespan. In most species – including all of our primate cousins – individuals continue breeding (or try to) until they drop dead. Humans do things differently. Unlike any of the other great-ape species, we have almost no overlap in breeding careers between mothers and daughters. Instead, the period when daughters become reproductively active coincides with the time when their mothers undergo a major physiological transition: menopause. Though we sometimes bemoan this period of life as signifying the start of old age, perhaps feeling as though we are becoming decrepit and defunct, I want to offer an alternative perspective. Menopause is an important switch-point in a woman's life that serves a specific purpose: this is when we change reproductive lanes, going from being breeders to being helpers.

The menopause is not simply an artefact of longer lifespans due to recent improvements in health care and lifestyles. Across most – if not all – societies, menopause occurs at around age fifty and is accompanied by an extended post-reproductive lifespan, even among people who lack access to technology or modern medicine. This includes contemporary hunter-gatherers and even historically high-mortality populations, such as plantation slaves living in Trinidad in the eighteenth century. Age of menopause is also heritable and, as more and more women living in industrialised societies delay having children until later in life, menopause also seems to be getting later.*

Taking a closer look at the underlying physiology shows us that menopause is not just part of the normal ageing process. Human females are born with around 2 million follicles *in situ*, each capable of producing an egg. This supply declines steadily throughout life: by the time she is twenty, the average female has around 100,000 follicles remaining, and around 50,000 by the time she is thirty-five. Extrapolating forwards, even with this rate of decline, the average woman *should* be able to continue reproducing until she is well into her sixties and maybe even seventies. But something strange happens when she is around thirty-eight years old. Now, the number of follicles nose-dives, going into a much steeper rate of decline. As a consequence, by the time she's fifty or thereabouts, her follicle levels drop below the minimum threshold required for regular menstrual cycles.

---

* Age at menopause is also influenced by environmental factors, like BMI and smoking. Interestingly, recent evidence finds that frequent sexual activity is associated with a later onset of menopause, suggesting that menopause might come earlier when the body 'knows' that there is no risk of becoming pregnant.

This highlights the mechanics of menopause. But it doesn't answer the question of *why*. Why do women experience this sharp, non-linear decrease in our fertility in our late thirties? And why do we then persist as sterile vessels, when it would seem that we have become reproductive dead ends?

*

To answer these questions, we need to take an evolutionary perspective. Through this lens, we come to realise that menopause is the outcome of an evolutionary battle, played out over millennia, between grandmothers and their daughters-in-law. Poor old mothers-in-law are the butt of a thousand schoolboy jokes but, as the saying goes, a joke is the truth wrapped in a smile.

The science isn't totally resolved on this, but there are good reasons to believe that dispersal in ancestral humans was female-biased. In other words, reproductive-age females tended to move to live with their 'husband' (I use this term loosely, to mean male breeding partner) and his family, rather than the other way around. An important consequence of female-biased dispersal is that the younger females (the wives) are potentially competing with their mothers-in-law over the limited resources needed to successfully raise children. We can make use of historical data sets of pre-industrial humans to get a feel for the effects of this competition. In Finland, the Lutheran church kept meticulous records of marriages, births and deaths from the 1700s until the early 1950s. Originally kept for tax purposes, these records now have a far greater value in helping us understand how selection might have acted upon historical human populations, in a time before inventions like the contraceptive pill and modern medicine could complicate estimates of fitness too much. This data shows that when a grandmother bred alongside her daughter-in-law, all of the children suffered. The costs were heavy: children were less than half as likely to survive to the age of fifteen when there was competition between breeding females. Nevertheless, co-breeding was also exceedingly uncommon, with just thirty or so grandmothers (out of more than 500) being reproductively active at the same time as their daughters-in-law. In most cases, we see a case of what looks like altruism: the older females concede to the younger ones in these reproductive battles. But how might grandmothers possibly benefit from curtailing their own reproduction and allowing younger females to breed unhindered?

This puzzle can be solved by considering the ways in which the younger and older females are related to one another's offspring. The mother-in-law has a vested genetic interest in any children produced by her son's wife (so long as they *are* definitely his children*). The wife, on the other hand, has no genetic interest whatsoever in any offspring produced by her mother-in-law. This is what's known as a relatedness asymmetry – and it weakens the mother-in-law's hand. A grandmother is disincentivised to breed, if doing so harms her grandchildren. The reverse is not true: the younger female's genes simply don't care about any costs they might impose on those residing in the mother-in-law and her children. As a consequence of this relatedness asymmetry, the grandmother is more likely to concede in any battle over reproduction; her pay-off coming, instead, in the form of grandchildren. Once she is committed physiologically to sterility, she can make the best of it by helping to raise her grandchildren. The benefits that grandmothers confer are well documented and can provide the selective impetus needed to favour the increased post-reproductive lifespan. From the ashes of an evolutionary conflict, grandmothers rise up.

* Probably quite a safe bet in the Finnish data set, where monogamous marriage was strictly enforced and adultery was severely punished. On average, the median estimate for misattributed paternity worldwide is low, estimated to be about 1–2% – the Finnish data set is likely to be comparable.

When all we have to go on are records of births, deaths and marriages, it is very difficult to infer how, exactly, grandmothers helped their grandchildren to survive. It is likely that these ancient grandmothers acted as repositories of knowledge, passing vital information on everything from breastfeeding to dealing with infants' illnesses. In some cultures, grandmothers breastfeed their grandchildren, and are able to produce milk for the child even when their own breeding attempts ended many years earlier. Grandmothers are also an extra pair of hands, someone who is available to help care for any dependent children, which allows mothers to undertake other jobs (foraging and paid labour, for example) that contribute to infant survival.

*

Another broad pattern emerging from these data sets is that not all grandmothers are equal: data from twenty-six historical and contemporary natural fertility populations has shown that maternal, rather than paternal, grandmothers make the most difference to the survival of their grandchildren. This is a bit confusing – we know women tended to have their babies in their husband's household, which might seem to imply that it would be the paternal grandmothers who would be doing the heavy lifting when it came to the childcare. The answer to this evolutionary riddle comes from yet another church database – Canadian this time – documenting the lives of French settlers in and around Quebec in the seventeenth and eighteenth centuries. The data shows that, even when daughters leave home to have their babies, the maternal grandmothers are still able to help out – so long as the daughters don't move too far away. Increased distance between mothers and their daughters corresponded with decreased survivorship of the daughter's offspring, likely because the maternal grandmother was less able to help out from afar. The general pattern therefore seems to be that conflict among mothers-in-law and daughters-in-law explains the evolution of menopause, but post-reproductive females direct their investments to the grandchildren they are most certain that they are related to: to their daughter's rather than to their son's children.

If women's post-reproductive lifespan has been extended by the benefits they could bestow upon their grandchildren, we might ask why grandmothers don't live for even longer. In fact, why do they have to die at all? Before answering this question, it is important to dispel the seemingly intuitive explanation that people die because they become old and decrepit. Senescence – the process of ageing – is not just a biological inevitability. Instead, it is something that is under the control of natural selection. If there were sufficient fitness advantage to living a bit longer and not being riddled with the ailments of old age, then we probably *would* have longer healthy lifespans. Ageing is what happens when evolution no longer sees a future for us, and selection becomes less assiduous in maintaining and overseeing basic physiological processes, like cell division. There's no point proofreading a document that no one is going to read.

So why don't grandmothers live forever? A recent analysis of the same Finnish data set indicates that grandmothers are only useful (in an evolutionary sense) for the child's first few years of life. Most of the children a grandmother can expect to help with will have been born by the time she is about seventy-five. Beyond this point, not only are grandmothers unhelpful for child survival, but they actually become a liability: living with one means that any child is less likely to survive into adulthood. This detrimental effect acts as a counterweight to the selective force favouring increased lifespan: in the end, grandmothers are no longer selected to live, but to die.

*

At this point, you might be wondering about men. There is some evidence that men experience an age-related decline in testosterone and are less likely to successfully attract female mating partners in old age – but men don't undergo a distinct menopause in the same binary fashion as women. So why do grandfathers get to keep their reproductive options open when grandmothers don't? What role do grandfathers play in the survival of their grandchildren? This is a burgeoning area of research and one where there is little real consensus, yet. A general pattern is that the presence of a grandfather is, on average, less decisive for the long-term survival of his grandchildren than is the presence of a grandmother (although there are a handful of studies which report exceptions to this general rule). If we take this as read for now, then the fact that men live so long starts to look a bit puzzling. In a monogamous society, a man's reproductive career ought to end when his wife hits the menopause. If grandfathers aren't especially pivotal for the survival of their grandchildren, then what keeps them alive beyond this age?

One possibility is that men might live longer because they can still find new breeding partners, even when their wives undergo menopause. In the Finnish data set, although divorce was not permitted, people could remarry if their current spouse died. Widowed men were three times as likely to remarry, compared to widowed women and, when they did so, they invariably took a much younger wife. More than 90% of the remarried men subsequently had more children, unlike the remarried women. Cross-cultural studies also suggest that, at least in some societies, a male's reproductive market value remains reasonably high, even as he enters his autumn and winter years, especially when men can accumulate wealth during their lives (for example in pastoralist and horticulturalist societies, like the Turkana and the Tsimane). In such societies, men can attract younger wives and continue having children when they are well into their sixties and even their seventies. The fact that men can reproduce at relatively low cost, and can potentially continue to accrue fitness benefits by having their own children late into life, might therefore explain why men also live long lives and don't go through a physiological 'andropause'.

*

So, our social life has shaped not just our physiology, but even our lifespan. In some of the most highly social species, selection seems to have stopped the clock on ageing entirely. The queens in social termite and ant colonies commonly live for over ten years, around a hundred times longer than their workers, despite spewing out hundreds or even thousands of eggs per day. To put this in perspective, in human terms, this would imply living to be *7,000* years old instead of a mere seventy.

Another set of creatures that seem to live extraordinarily long lives are the social mole-rats. By anyone's standards, the mole-rat is an odd-looking specimen: a squinting little creature, fronted with a pair of snapping, yellowed incisors that protrude from the outside of its closed lips. Although Damaraland mole-rats are rendered marginally more appealing by virtue of having some fur, their naked mole-rat cousin rather resembles a human penis with teeth – to the extent that a 2017 campaign run by the Canadian Centre for Child Protection, trying to prevent teenage boys from texting pictures of their genitals to others, suggested that they might send pictures of these 'long, veiny and fleshy' animals instead. The first person to catch a naked mole-rat was so appalled by its appearance that he concluded that he must have caught a diseased individual from another species, a misapprehension that took another thirteen years to remedy.

Despite their strange appearance, mole-rats are among the most fascinating and well-studied animals on the planet. I even studied them myself for a short time – a study which was aborted due to insufficient space to house the giant Perspex colony structures we needed, and (I'll admit it) my ineptitude and slight fear of handling mole-rats. To grab one without losing a finger requires the handler to swiftly grasp the mole-rat by the tail, a little nub that protrudes no more than a few millimetres from its rear end. Mole-rats may have poor vision but they have amazing perception of air movements (that indicate, for example, an approaching hand) and the ability to snap their neck backwards at an extreme angle to latch their teeth onto the aerial threat.

Researchers braver than me have worked extensively on the two highly social species in this family: the Damaraland and the naked mole-rats – these are the rodents that mistook themselves for ants. Like many ants, these two species of mole-rats live in extensive subterranean societies. Damaraland mole-rat colonies are quite small, numbering fewer than twenty individuals; but naked-mole-rat colonies can be hundreds strong. Each colony is ruled by a single breeding queen and, unlike ants, the mole-rats also have a king. Mole-rat workers are not permanently sterile, but, so long as the queen is present and there are no unrelated males in the group, the subordinate females are physiologically suppressed and typically do not ovulate – a bit like women who take the hormonal contraceptive pill. As a consequence, the majority of mole-rat workers never breed during their lives. Like ant queens, when a female mole-rat inherits the crown, she embarks upon a renewed period of growth. She doesn't enlarge her skull and cranial musculature (digging tunnels with the teeth is, after all, a worker's job) but instead selectively elongates her body. In doing so, she expands her abdominal cavity, meaning that she that can gestate larger litters and give birth to heavier pups: she becomes specialised for reproduction. Like the insect queens, the breeding females in mole-rat societies also have extended lifespans, living for up to thirty years, in contrast to their workers who live for around eight years.

Have these (near-)immortal queens unlocked the elixir of eternal youth – and if so, how have they done it? To answer this, it helps to think about ageing as being more of a design feature than a design flaw. Life inevitably involves trade-offs because resources are limited. If you only have a fixed amount of money in the bank, anything you spend on reproduction will not be available to spend on maintenance and repair of the body. As such, a near-universal law is that increased efforts in reproduction come with the cost of decreased survival. Ageing stems from resource-allocation decisions about whether to front-load reproductive efforts, or to spend a bit more on early growth and survival and delaying reproduction until after you've made some of these initial investments.

The ageless queens of these highly social species seem to flout this otherwise universal law: data from the ants indicates that they can reproduce at fearsome rates without suffering any cost in terms of ageing, and the mole-rat queens live longer than their workers. Long queen lifespans are not just a function of these individuals being higher-quality – and thus starting out with more of the metaphorical money in the bank which would allow them to reproduce *and* live longer than the workers. Experiments in ants where the queen is forced to lay more eggs (by removing the ones that she already laid) show that this has no effect on her lifespan whatsoever. A creature that seems to incur no mortality cost associated with reproduction is known as a 'Darwinian demon', so named because they have mysteriously worked out how to have their reproductive cake and eat it too. How do they do it?

*

One way to get evolution to give you a long lifespan is to be less likely to die. I realise this sounds tautological – like the punchline to a naff joke ('I could have lived longer if I hadn't died first') but there is more to it than this. The risk of extrinsic mortality (death due to predation or disease, for example) affects the evolution of lifespan, by shaping the physiological process of ageing. Think about a fruit fly. Fruit flies don't have long life expectancies because they face a pretty high risk of being eaten or squashed by other creatures. As a consequence, fruit flies are adapted to front-load reproductive efforts, despite these same adaptations carrying deleterious consequences later in life. Even when housed in safe laboratory conditions, fruit flies cannot typically survive more than a few weeks. For species that can expect to live longer, selection will act more strongly to delay the process of ageing and, in turn, give rise to longer potential lifespans.

One of the most effective ways for a species to increase its lifespan is to be good at evading predators, for example by being able to fly. Bats have lifespans around 3.5 times that of similar-sized mammals – the tiny Brandt's bat is barely as long as your finger and yet can live to be over forty years old. Birds also have longer lifespans than would be expected based on body size alone; and flighted birds typically outlive similar-sized flightless species. Flight is linked to lifespan because it offers an alternative escape route, thereby reducing predation. For the same reason, tree-dwelling (arboreal) species also tend to have longer lifespans than their terrestrial counterparts.

This starts to shed some light on the extraordinary lifespans of naked mole-rats and eusocial insect queens. In part, their increased lifespan also reflects a protected way of life: the ants with the longest lifespans tend to live in highly protected nests, while mole-rats live deep underground, shielded from predators and the fluctuations in temperature that occur above their heads. Ant queens are spared from performing dangerous tasks outside the colony – workers do all of the foraging. Ants also structure their society so that foragers seldom come into contact with the queen. The foraging individuals are most exposed to disease and infection from the outside world and thus don't mingle with the workers who tend the brood or the queen, who are hidden away deep inside the nest's internal cavities. I like to imagine these privileged workers as being a bit like the intensive care team working inside a hospital, with the queen and the brood representing their vulnerable patients. Experimentally infecting foragers with a pathogen prompts the ant intensive care team to sequester the valuable brood even deeper inside the nest, further isolating this section of the colony from the others who might infect it.

These safety mechanisms help protect the queen but they are not the only reason she has a long life. We've already seen that ant colonies can be viewed as super-organisms, where the majority of sterile workers are equivalent to the cells that build a multicellular body. This means that a queen ant has two distinct phases that define her reproductive career: the growth phase (where she produces sterile workers), and the 'true' reproductive phase (where she produces reproductive males and new potential queens). According to this view, the early stage of colony growth is akin to the juvenile period that most animals experience, where they grow from small juveniles into larger adults. The true reproductive phase of the ant colony corresponds to the period when an animal switches its focus from growth to reproduction. It is in the production of sexual males and new queens that a queen ant reaches adulthood, and buys her genetic lottery tickets to the next generation.

*

Before we move on, let's pause for a moment to take stock of the bigger picture. By now, we are starting to appreciate the myriad ways that cooperation touches and moulds our lives. It is not just about what we *do*, but who and what we *are*. Cooperation has shaped our reproductive physiology, making grandmothers out of some of us. It helps to keep us alive, both in terms of the help we might receive and via its effects on our lifespan.

Cooperation has taken centre stage in this story, but we should not forget about conflict, which is waiting in the wings. In the next chapter, we are going to see how these two forces interact and how ostensibly loving families can also be hotbeds of rivalry and menace.

9
ASCENDING THE THRONE

> Me against my brother, me and my brother against my cousin, me and my cousin against the stranger.
>
> Arab proverb

As anyone with relatives is aware, families aren't always harmonious. As a child, I remember learning the disturbing story of the two princes, Edward and Richard, only sons of Edward IV. Following their father's death in 1483, the two boys, aged twelve and nine, were left in the custody of their uncle, Richard. In hindsight, this was a bad decision. Richard locked the boys in the Tower of London, ostensibly on the grounds that this was the safest place for them to stay while the older of the two princes (Edward) awaited his coronation. But the boys disappeared and their uncle, who was next in line, took the crown himself. Nearly 200 years later, the remains of two small humans were found in the tower, a discovery which implicated long-dead Uncle Richard in the disappearance and possible murder of his young nephews.

Stories of familial treachery repeat themselves throughout history, starting with the Good Book itself and continuing to the present day. The Bible makes a strong opening move in the sibling-rivalry genre, with the news that Cain kills his brother Abel spilling onto the page in the first act. The reader barely has time to digest the news that Eve is pregnant with her firstborn son, before being confronted with his murderous deed a few sentences later. The North Korean dictator, Kim Jong-un is thought to have ordered the assassination of his older half-brother at an airport in 2017. Indeed, North Korea takes the potential for sibling rivalry so seriously that successors to the throne are not even raised together: it is thought that Kim Jong-un first met his half-brother just five years before the latter was murdered.

*

But family strife is not an exclusively human foible. Sand tiger sharks hatch inside their mother's womb and embark on a feeding spree, cannibalising their siblings *in utero*. Even in cooperatively breeding species, where societies seem to be built on tolerance and help, we see frequent squabbles and tussles. In most vertebrate species, helpers are hopeful breeders that anticipate inheriting a breeding position one day. Most of the time, subordinates are prevented from co-breeding alongside the dominant pair because they are surrounded by relatives. However, if the opposite-sex dominant individual is a step-parent – or if subordinates can chance a fleeting encounter with an unrelated visitor – then conflict begins to bare its teeth. For example, subordinate female meerkats won't mate with the dominant male in the group if he is their father, but they can meet potential suitors in other ways. Subordinate males – who are similarly prevented from breeding at home – make brief forays into neighbouring territories, trying to mate with unrelated females when they do so. These illicit encounters allow opportunities for subordinates of both sexes to advance their own direct fitness interests, by breeding – even if they lack unrelated breeding partners in their own group.

A dominant female meerkat may not be able to prevent her subordinates from getting pregnant, but she has other tricks up her sleeve to make sure that her sister's (or daughter's) attempts to breed do not interfere with her own. Dominant females routinely evict pregnant females, by aggressively chasing them from the group.* Forced to live alone, the subordinate female rapidly loses weight and often aborts any litter she is carrying. When a subordinate *does* manage to give birth to her pups in the group, the dominant female will usually kill and eat them – making meerkats serious contenders for the 'World's Worst Grandmothers' award. The dominant female can afford to pursue this infanticidal strategy so long as her own pups have not yet been born. However, if the dominant female has already given birth, then the subordinate's pups will usually be safe (albeit at a competitive disadvantage) because the dominant – who cannot easily discern her own pups from anyone else's – will not risk accidentally devouring her own offspring.

Other species find different solutions to the problem of who gets to breed. Banded mongooses live in large societies, like meerkats, but in relatively richer environments. Unlike meerkats, where just one female monopolises reproduction, most females in a banded mongoose group do breed. However, there is still the risk that females might try to thwart one another's breeding efforts by cannibalising pups that are not their own – and it is the gang of dominant females in the pack that are the most likely to kill and eat the pups born to subordinates. This was cleverly demonstrated in a study led by Professor Mike Cant and colleagues, which involved putting banded mongoose females on the contraceptive pill. Females that were physiologically suppressed did not give birth to pups, and could therefore identify that any pups that were born in the group as being 'not mine'. Whether these pups were killed or not depended on which females were given the pill. When subordinates were prevented from breeding, most pups still survived, indicating that the

---

* Evictions are more common when food is plentiful: a dominant female cannot risk losing a helper when times are tough.

subordinates were not killing the offspring born to the dominant females. But when the dominants were on the pill, they routinely killed the pups born to the subordinate females.

Under more natural conditions, banded mongooses get around this problem by perfectly coordinating their breeding activities: all of the females in the group give birth on exactly the same day. When no female can be entirely sure which pups are hers, there is a much larger disincentive to kill offspring. As you might expect, the synchronisation of birthing is driven by subordinate females: although they tend to mate a few days later than the dominant females, they seem to speed up their pregnancies so as to synchronise births. Better to have your pups a few days early than to risk them being eaten by their female relatives.

*

In and of itself, high relatedness is not, therefore, a panacea for fostering cooperation. If individuals can advance their own interests by competing with, or even killing, their relatives, they often do so. Squabbles within groups are often detrimental to group efficiency and building the most cooperative societies therefore involves suppressing the conflicts that occur between group members. In most vertebrate societies, the task of keeping the workers in line falls to the dominants. But there's an obvious problem with this solution: there is a limit on how many workers a single dominant is able to effectively control.

One particularly elegant demonstration of this fact comes from a study of Colombian *Polistes* wasps, carried out by Mary Jane West-Eberhard in the 1980s. *Polistes* wasp society is structured a little like those of the babblers and meerkats we've already met: a single queen does most of the breeding, but she is flanked by hopeful subordinates who will lay eggs if they get the chance. Ordinarily, the queen polices the behaviour of her subordinates through a despotic regime of aggression and intimidation. But, in her experiment, West-Eberhard prevented the queen from harassing the subordinates by tying a piece of nylon around the queen's waist and tethering her to a branch at one end of the nest. It is all too tempting to imagine the fury of the newly impotent queen, straining at the end of a piece of string and quite unable to police the activities of the emboldened subordinates at the other end of the nest. As you might expect, the subordinates laid eggs in the part of the nest where the queen couldn't reach or harass them. Fortunately for the queen in this experiment, the nylon tether broke on the fifth day, freeing her to roam over the entire nest once again and to resume her physical harassment of the subordinate females. Social order was quickly re-established. Nevertheless, the experiment demonstrates the limitations of authority when it is wielded by a single individual: the *Polistes* queen could only police the subordinates when she could physically reach them.

When the actions of group members are policed by a single individual in this way, it places a limit on the size that a cooperative group can achieve before collapsing under the weight of individual self-interest. Groups become societies only when the members begin to police themselves. In multicellular bodies, there is no centralised headquarters that polices the actions of all the individual genes and cells. Instead, remember that the genes (and cells) in your body form a democratic parliament of sorts, working together to suppress the interests of selfish elements. It turns out that some of the eusocial insects also function more like a multicellular body in this respect, where the potentially selfish actions of the workers are policed not by the queen, but by the other workers in the colony.

Until now, we have thought of the workers in most ant societies as being sterile castes – and in some senses this is true. Most ant workers are not capable of sexual reproduction. But among the Hymenoptera (ants, bees and wasps) a male can be produced from an unfertilised egg. The proper scientific name for this kind of sex determination is 'haplodiploidy' meaning, quite literally, that males are haploid (they have half of the full complement of chromosomes) and females are diploid. This results in some curious outcomes: males have mothers, but not fathers – and can themselves only produce daughters and not sons. Haplodiploidy has two important consequences that affect conflict within a social insect colony: female workers can produce males by laying unfertilised eggs, and workers are more related to their sisters than they are to their brothers. The fact that workers can lay their own eggs means that they come into conflict with the queen over reproduction, and the relatedness asymmetry means that they will, at some point, try to kill her.

To understand how haplodiploidy affects worker relatedness to other members of their colony, let's assume the simplest case (for now) of a queen that has mated with just one male,* meaning that her daughters are full (not half) siblings. In Hymenoptera, males pump their entire genome into each sperm cell, meaning that all his daughters inherit the same genome from him. This half of a female's genome is therefore identical to that of her sisters. The other half of the genome comes from the queen: here, females have a 50% chance on average of sharing any given gene with a sister. The average female worker can therefore expect to share around three-quarters of her genes with any sister: they have a 100% chance of inheriting any given gene from their father, and a 50% chance of inheriting any given gene from their mother. This relatedness is high, when compared with the conventional 50% for full siblings in normal diploid species, like us. A female worker is far less related to her brothers, however. Remember that males don't carry any paternal genes, so a female does not share any genes from her father with her brother. As above, any gene inherited from the mother will also be present in the brother with a 50% probability. A female worker therefore

* In many insects, a single ejaculate from one male can be stored by the queen for years, and used to fertilise many eggs.

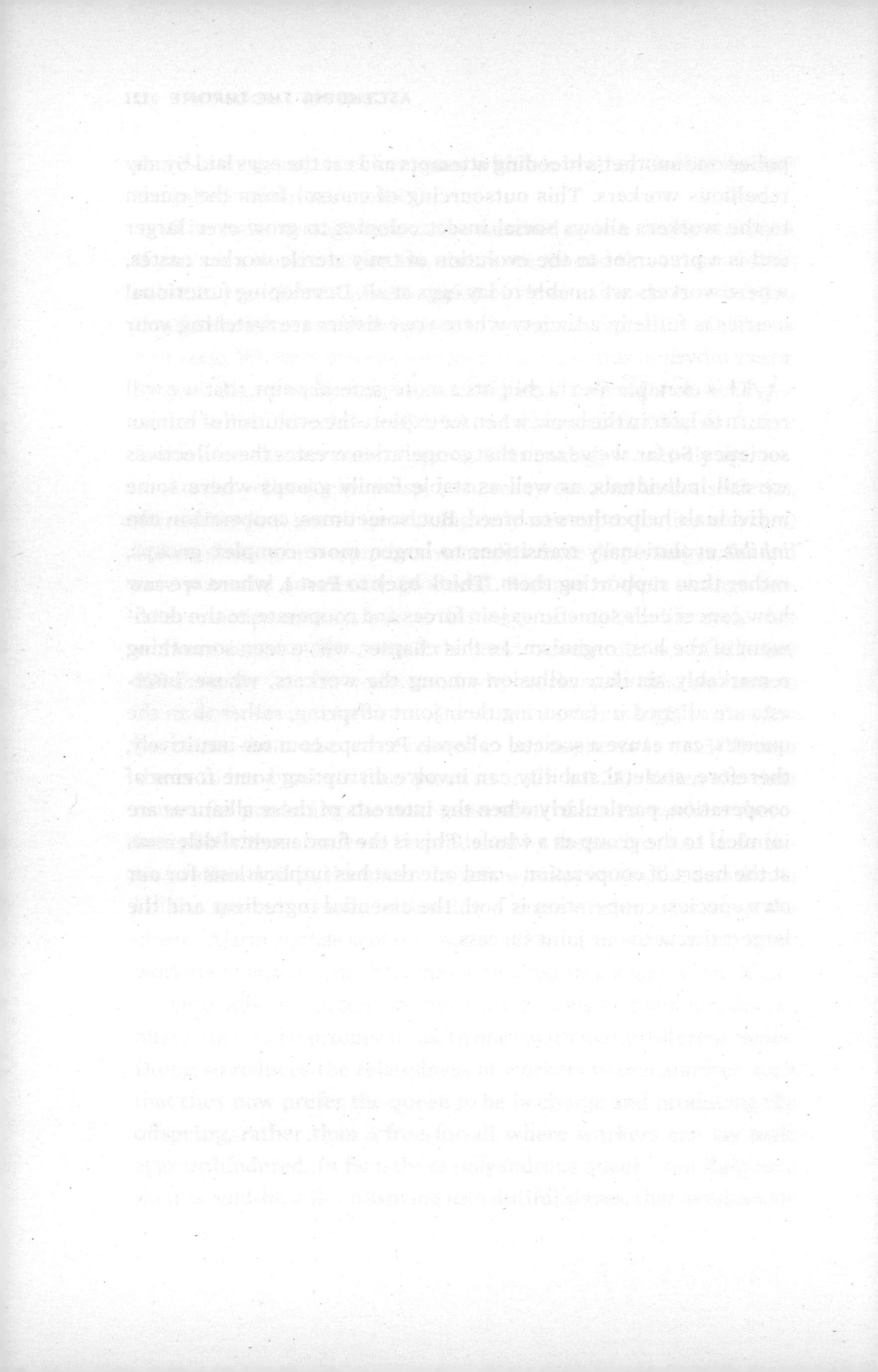

policed the matriarch's breeding attempts and eat the eggs laid by any rebellious workers. This outsourcing of control from the queen to the workers allows eusocial insect colonies to grow ever larger and is a precursor to the evolution of truly sterile worker castes, where workers are unable to lay eggs at all. Developing functional ovaries is futile in a society where your sisters are watching your [illegible]

[illegible] example [illegible] point that we will return to later in the book, when we explore the evolution of human societies. So far, we've seen that cooperation creates the collectives we call individuals, as well as stable family groups where some individuals help others to breed. But sometimes, cooperation can inhibit evolutionary transitions to larger, more complex groups, rather than supporting them. Think back to Part 1, where we saw how cancer cells sometimes join forces and cooperate, to the detriment of the host organism. In this chapter, we've seen something remarkably similar: collusion among the workers, whose interests are aligned in favouring their joint offspring, rather than the queen's, can cause a societal collapse. Perhaps counter-intuitively, therefore, societal stability can involve disrupting some forms of cooperation, particularly when the interests of these alliances are inimical to the group as a whole. This is the fundamental [illegible] at the heart of cooperation – and one that has implications for our [illegible] cooperation [illegible]. The essential ingredient and the [illegible]

Part 3

WIDENING THE NET

In 1987, an eighteen-month-old toddler called Jessica McClure fell twenty-two feet into a well in her aunt's garden in Texas and became trapped there for fifty-six hours. Her rescue involved the fire and the police departments and received round-the-clock coverage from various national news channels. By the time she was rescued from the well, Jessica had received over $800,000 in donations from members of the public, people who were moved to donate out of a concern for the toddler's plight and a sense of empathy.

Nothing like this ever happens in non-human species – but why not? There are the obvious differences, of course. No other species has institutions that are dedicated to helping people, like the fire and police departments. Ants don't have news channels and chimpanzees don't use money. But we differ from other species in more fundamental ways. We routinely and voluntarily help individuals that we are not related to, whom we never expect to meet again, and where there is nothing material to be gained from the interaction. And the weight of evidence also suggests that we may be the only species on the planet with the ability to feel another person's pain – and to be motivated to alleviate this if we can.

It is easy to imagine that these traits have evolved because of our history as cooperative breeders: kindness, tolerance and willingness to help others all seem like character traits that would be helpful in the family setting. But this can't be all there is to it. The cast of family-living species we have met may be a diverse bunch, but they have one important thing in common: they are driven by nepotism. Whether we consider ants, babblers, bees, or mole-rats, helping is ringfenced within the family group. Humans, on the other hand, widen the net, extending help to those beyond the family and even to complete strangers. It is also telling that some of the most important socio-cognitive traits that set humans apart from other species – a concern for the welfare of others, the ability to take another person's perspective and to understand and share their mental states – are traits that are conspicuously lacking among the other cooperatively breeding species on the planet. Meerkat helpers generously donate food to hungry pups, but they aren't the slightest bit concerned with the recipient's well-being. Ants rescue nest-mates wounded in battle, but there is not a jot of evidence that such behaviour is driven by empathy. This seems to present a bit of a paradox. Humans are one of the most cooperative species on the planet, a trait we share with other cooperatively breeding species. But our version of sociality is built on different cognitive foundations. What is it that sets us apart from the other species on the planet? And how did we get here?

In what follows, we are going to explore the foundations of the large-scale cooperation that characterises human societies. We will have to start small, by understanding how and why evolution would ever favour cooperation outside of family groups. We aren't the only species to help strangers, but we do it on a scale that is simply unparalleled in the natural world. Part of the reason we are so successful in this domain is that we have more powerful cognitive abilities at our disposal, abilities that allow us to see beyond the constraints of the scenarios that nature confronts us with and to envisage that other, more cooperative worlds are possible. This power of imagination allows us to invent new rules for our social interactions – rules that help us to coordinate our actions so that we can avoid conflict, to achieve outcomes that are greater than the sum of our parts, and to incentivise cooperation where it might otherwise fail. Our ability to create and change the rules of the game of life is the key factor that has allowed us to scale up cooperation, from investments in a few important relationships to a world where we routinely and instinctively trust and help others, even those we have never met and might never meet again.

10
THE SOCIAL DILEMMA

If men were angels, no government would be necessary.
James Madison, *The Federalist Papers*, 1788

'I'd be very, very happy to go home with £50,000.' Stephen gesticulates to the studio audience. 'If I stole off you, every single person up there would come over here and lynch me.' He leans over the small table separating them, taking her hand inside both of his, and lowers his voice, 'I can tell you, I am going to split.'

Stephen and Sarah are two strangers playing the most human of games: trying to figure out whether they can trust one another. This is the finale of *Golden Balls* – a British TV take on the classic Prisoner's Dilemma game. Stephen and Sarah each have to decide whether to play nice or not, while at the same time trying to work out what the other is likely to do. With £100,000 at stake, there is a huge incentive to get it right.

After a few moments of conversation, Stephen and Sarah will reveal their decisions to one another by holding up a golden ball with the word SPLIT or STEAL inscribed inside it. If both players choose to split the money, they will go home with £50,000 each. But if one player chooses steal, while the other splits, then the stealer takes the whole lot. If they both choose steal, they both get nothing.

With tears in her eyes, Sarah pleads with Stephen to trust her, 'Everyone who knew me would just be disgusted if I stole'. The tension hangs in the air as the host reminds the two players to check the balls once more before making their final decision. Upon the big reveal, we see that Stephen, true to his word, chose to split the money, whereas Sarah holds up the ball declaring her intent to steal. She takes the entire £100,000, leaving Stephen with nothing.

*

Although it is tempting to malign Sarah as a selfish, perhaps even immoral individual, if we consider the pay-offs involved in a standard Prisoner's Dilemma game, we might start to wonder why anybody would ever play nice in a scenario where the incentives seem to be stacked against doing so. Regardless of whether we are talking about pay-offs in terms of money or food or offspring, and whether the amounts at stake are minuscule or gargantuan, we can classify any two player interaction as a Prisoner's Dilemma if the best mutual outcome is for both parties to cooperate, but each individual can gain a short-term advantage from defecting.

We face many scenarios that appear to have this sort of incentive structure in real life – whether we are interacting with just one other person, or in a much larger group. Sometimes these interactions are with people we know and like, and at other times they are with complete strangers, including people we might never even meet. Should I do the washing up, or let other members of the household do it for me? Should I ride my bike, or drive my car to work? Should I get my child vaccinated against infectious diseases, or rely on the herd immunity that results from other people vaccinating their children instead? Broadly speaking, the kinds of cooperation problems we encounter from day to day can be summarised under one common header: social dilemmas. They are *social* because our decisions affect other people (even if this is not always obvious). And they are *dilemmas* because individual and collective interests diverge. These incentive structures, which pit individual against collective welfare, don't paint a particularly optimistic picture for the emergence of cooperation. And yet, we often perform seemingly selfless acts that require us to incur immediate personal costs that deliver benefits to others. How can we make sense of this puzzling propensity?

*

Asking this question can be seen as unnecessarily cynical. People will roll their eyes and tut, explaining that they cooperate because they have a genuine concern for the beneficiary, or because they're a nice person, or because it would be immoral to cheat. I'm not denying the existence of such motives. Far from it: we can even study and quantify them. Early psychology experiments showed that people can be sufficiently motivated by empathic concern to offer to receive painful electric shocks, rather than watching the shocks being delivered to a 'stranger' (actually an actor) instead. We frequently even enjoy helping others: that fuzzy feeling you get when you do a nice thing for somebody is something that economists call the 'warm glow of giving' and, again, it's something we can study. If you place people in a brain scanner and give them the opportunity to send money to a good cause, you will see the brain areas associated with reward fire up. These are the same areas that are active when people eat delicious food, or have sex, or indulge in recreational drugs like nicotine or cocaine. Another study where researchers gave participants a wad of cash and instructed them to either spend it on themselves or on another person found that, in the evening, the people who had spent their money on others rather than themselves felt happier. The same team of researchers have also shown that toddlers smile more when instructed to share a cracker with a toy puppet (compared to when no sharing is requested), and that spending money on others can even help to lower blood pressure and improve cardiovascular health.

As interesting as these kinds of experiments are, the explanations they offer don't really get us closer to the deeper answer of why people help others. Asserting that people are kind 'because it's in their genes' or 'because they feel empathy' or 'because they enjoy helping others', is a sleight of hand; we are simply replacing one kind of question with another. Thinking in evolutionary terms poses a more fundamental question: why do we have a brain full of psychological machinery that drives us to perform such costly behaviours? It is not hard to envisage a world where individuals whose brains *aren't* designed to feel concern for others or to experience the warm glow of giving or to behave morally, might outperform those of a more altruistic disposition because these hyper-rational, self-interested types would not incur the costs associated with cooperating.

What's more, although humans seem to rely on them, fancy psychological mechanisms like empathy and morality are not even needed to promote cooperation. As we shall see, many non-human species find cooperative solutions to social dilemmas even when they do not have a sense of morality or a capacity for empathy. In some cases, these sophisticated capacities can even do more harm than good. The Jessica McClure case highlights a quirk of human psychology known as the 'identifiable victim effect', whereby someone can feel a visceral urge to help when faced with the plight of a single individual, but is unfazed by the suffering of thousands. Human empathy is bounded and fickle: it is only bestowed upon 'worthy' individuals, and we too often turn a blind eye in situations where help is most needed. We certainly cannot invoke empathic concern or a sense of morality as a comprehensive answer to the question of why helpful behaviour exists.

*

To understand why individuals are prepared to pay costs to help other individuals on a more fundamental level, we need to think about how a tendency to invest in costly helping behaviour could have been favoured by selection. Before we think about this, however, it is important that we don't get bogged down in a misunderstanding. I am sometimes accused of 'taking the altruism out of altruism', by showing how apparent acts of kindness can be reconciled with long-term benefits to the helpful person. Someone who gives to charity might bristle at the suggestion that there can be a downstream benefit to their selfless actions because this implies that their generosity is calculated and knowingly self-serving. This can grate against what many of us experience as an altruistically pure motive to help someone in need.

I understand this consternation, but it is based on a misconception. Answers to *why* questions can come in many flavours, and evolutionary biologists are interested in two of these in particular. We call them proximate and ultimate explanations. Proximate-level answers to the question of why people help others concern the *immediate* causes of the behaviour. These explanations might include appeals to context ('I don't mind helping her out; she's my friend'); to personality ('Joe always thinks of others first; he's a really kind person'); or to empathic concern ('She felt so sorry for the man shivering on the pavement outside that she brought him a cup of tea from the station cafe'). Proximate-level explanations also include things we can't so easily put our finger on like hormonal causes (remember the way that testosterone is implicated in paternal care); or the size of different brain structures or patterns of neuronal activity. For instance, a recent study that examined brains of extreme altruists (people who had donated a kidney to someone else) compared to a control group found differences in the structure and the functioning of the altruists' brains. Regions of the brain thought to be involved in producing empathic responses to others were enlarged and more readily activated in the altruist group. Interestingly, these same brain regions were smaller and less empathically active in people with a tendency towards psychopathic traits.

When we look for an ultimate explanation, we are after something different: we want to know how a propensity for helping others might have been favoured by selection. Ultimate explanations help us to understand *why* we feel the motives we do, and *why* brains are designed the way they are. But these explanations are sometimes difficult to accept because we are often unaware of the adaptive significance of our actions. An individual need not know anything about the evolutionary consequences of a trait for the trait to be under positive selection.

To grasp this, a completely different example might help. If you are lucky enough to witness a lioness hunting, you might ask why she chases the impala. The proximate explanation is (probably) that she is hungry, or maybe that she has cubs to feed. The ultimate explanation is that a lioness that hunts impala will tend to have more offspring than one who sits under a tree all day. Hunger is therefore an evolved psychological mechanism – that manifests as an urge. This proximate mechanism makes the lioness do something that is key to her survival and reproductive success: hunt and eat prey. Sex is another good example. The psychological impetus and the evolutionary rationale for having sex need not coincide: we don't necessarily desire or expect offspring to result from every (or even any) sexual encounter. Nevertheless, this doesn't change the fact that sex exists because it increases reproductive success.

The same goes for altruism. We can accept the possibility that altruism yields downstream benefits to the altruist, without also undermining a person's motives for performing the behaviour. Just as evolution shaped our psychology to give us feelings like hunger (a signal to eat) or to find sex pleasurable, it is also likely to have shaped the motives underlying helpful, kind or moral behaviour, thereby prompting us to enjoy doing what is in our genes' interests.

So, how might cooperative traits be in an individual's long-term interest? We've already seen that costly helping behaviours can sometimes be favoured when they yield benefits to relatives. But this isn't the only way for cooperation to thrive: sometimes help can be favoured by evolution because the helpful individuals themselves eventually derive a healthy return on their investments. Let's explore how this could happen.

*

The simplest mechanism that helps cooperation to thrive is when helpful individuals receive help in return. This is the principle of reciprocity, initially formalised in the 1970s by the evolutionary biologist Robert Trivers. Reciprocity is so fundamental for driving cooperation that it has become enshrined into well-known proverbs. *Quid pro quo. You scratch my back, I'll scratch yours. Do as you would be done by. One good deed deserves another*. These maxims exist in other languages too. In Italian, *una mano lava l'altra* translates into the particularly lovely 'one hand washes the other', a phrase that also exists in German (*ein Hand wäscht die andere*). In Spanish, *hoy por ti, mañana por mi* means, roughly, 'today for you, tomorrow for me'. In religious texts, the principle of reciprocity is known simply as the Golden Rule. Around the world, reciprocity seems to be a universal imperative to guide behaviour. The prospects for players locked in a Prisoner's Dilemma game start to look a little different if you imagine the game being played over multiple rounds, instead of just once. If you expect to interact repeatedly, and your partner is a reciprocator (who is prepared to cooperate so long as you do the same), then it can be profitable to cooperate yourself.

Reciprocity is not widespread in nature, but it has been independently discovered by a handful of species who, like humans, have a requirement for social exchange. For example, vampire bats need a blood meal each day if they are to survive, but finding this meal is sometimes tricky. These bats live in large colonies, and bats who have successfully foraged often regurgitate blood meals for unlucky roost-mates who weren't able to find food that night. A bat who donates blood one evening is likely to be repaid by the beneficiary another night. Other species exchange services rather than blood meals. For instance, the brightly coloured rabbitfish that live on coral reefs team up with a partner while foraging, and take turns to watch for predators. A particularly neat example of reciprocity comes from experiments performed with pied flycatchers, showing that these birds only help the neighbours who helped them previously. Although flycatchers breed in pairs, they frequently harass predators that appear in the vicinity of a neighbour's nest. This is a costly endeavour, as birds spend time and energy when they help their neighbours, as well as potentially exposing themselves to a predation risk by flying towards an area where a predator has been spotted. Researchers investigated this helping behaviour by placing a model owl predator at the nest of one flycatcher pair and preventing the neighbours from coming to help drive the predator away (by temporarily trapping the birds). They found that the focal birds refused to come to the assistance of their 'nasty' neighbours a few days later, when the owl was stationed at the defectors' nest instead.

Vampire bats and pied flycatchers can be relatively assured that they will interact with their social partners in the future, meaning that the logic of repeated interactions can favour reciprocity. This is not the case in other species, however. Take the hamlet fish. Here, partners interact just once in the high-stakes game of making babies. Hamlet fish are simultaneous hermaphrodites, which means that they have a rather unusual sex life: individuals produce both eggs and sperm at the same time. Despite having all the apparatus needed to make hamlet babies, these fish cannot self-fertilise. In order to procreate, they therefore need to find a willing partner with whom to exchange eggs and sperm. But when one hermaphroditic hamlet fish meets another, they face a situation very much like a Prisoner's Dilemma. Each fish wants access to the partner's eggs, but does not want to give up their own eggs too easily (the eggs being the costlier of the two gamete types to produce). In such a scenario, any fish who offers all their eggs in one move is taking a risk, since the partner could fertilise the eggs and then swim away without offering their own eggs in return.

Hamlet fish resolve this dilemma in two key ways. The first tactic is to parcel their investments of eggs into smaller bundles. Rather than releasing all of their eggs at once, each hamlet releases a few eggs and allows the partner to fertilise them, and then waits for the partner to release a few eggs of its own before continuing with the exchange. The old adage that you shouldn't put all your eggs in a single basket turns out to be based on evolutionary logic: in hamlet fish, this prudent strategy turns what would otherwise be a one-off interaction into a repeated exchange. The second tactic hamlet fish use is to start these proceedings in the early evening. Starting trading as night falls is important because the whole process must be completed before it gets too dark. Once a hamlet finds a willing partner, its best bet is therefore to play nice and release a batch of eggs when the partner does the same, rather than accepting one small parcel and swimming off in the hope of finding another gullible fish to exploit. Although such a strategy might be profitable if there was unlimited time, a hamlet who played a roving defector strategy risks not being able to find another partner to trade with before the light is lost, and trading ceases for the day.

This strategy of parcelling investments into smaller packages (such that it becomes more profitable to exchange with the partner you have than to try to find a new one) is quite common in nature. For instance, in many species, animals exchange grooming with one another, removing ticks and other parasites from hard-to-reach places. In these interactions, however, we seldom observe one individual fussing over its partner for a long stint and then waiting expectantly for the partner to return the favour. Such an audacious opening move would leave the investor open to exploitation. Rather, like the hamlet fish, grooming partners also parcel these bouts into smaller units of investment: one individual grooms the partner for a few moments, the partner then reciprocates, and so on. Even if the whole interaction ends up taking a long time, the exchangeable units of investment within the interaction are kept much smaller.

*

A core assumption of the Prisoner's Dilemma model is that individuals pay a cost to bestow a benefit on their partner. They should therefore try to avoid incurring this cost whenever possible, by defecting on a partner who defects, and exploiting a partner when they can get away with it. This is a nice theory – but, aside from the handful of examples I just gave, one that seems to bear no relation whatsoever to most of what we see in nature. The quintessential repeated interaction is a friendship and, in the context of these established relationships, individuals seem to take a more flexible approach to give and take. Rather than counting the pennies in every transaction, or insisting that it is your turn to buy the beer today because I bought it last week, friends seem to provide help or services with no strings attached, and do not retaliate against 'defecting' partners by withholding help in future interactions.

To understand why it might be advantageous to help someone else to whom you are not related, even if they don't always help you in return, let's use a slightly far-fetched vignette. Imagine that you're on a boat that has sprung a leak and the boat will rapidly sink unless you and your crew member work together to bail out the water. In a scenario like this, your interests become very much aligned with those of your partner: you both agree that working hard to bail the water is the best thing to do and neither is tempted to shirk his duties. But there is more to it than this: you now also have a stake in your partner's well-being and survival, to the extent that this affects his ability to keep bailing water out of the boat. If you find a chocolate bar in your pocket, it might be prudent to share this, if it provides the necessary sustenance for your partner to keep working towards the common goal. In sharing the chocolate bar, you are not expecting your partner to return the favour but instead are allowing him to continue with the task that benefits him, because his labour benefits you as a by-product.

Occasionally, your own survival might depend upon your partner's existence, even if he does nothing to help you or a common cause. To continue with the sinking-boat example, if your bailing efforts fail and you are forced to try to swim for shore, your chances of being eaten by a shark are halved if you are swimming alongside another human, compared to when you go it alone.

This example is obviously silly but illustrates a hugely important theme in the evolution of social behaviour: *interdependence*. This is the notion that individuals can sometimes have a stake in their partner's fitness, such that it can pay to invest in the partner even if the partner will not help in return. For instance, group size really does affect individual survival and reproductive success in many ways (by diluting predation risk or increasing success in inter-group conflict, for example) meaning that group-living species are always likely to be interdependent upon one another to some extent. Interdependence can explain why cooperation thrives, even in circumstances where reciprocity does not seem to operate.

Many of the examples of cooperation in nature that were initially interpreted as reciprocal interactions occurring within the context of Prisoner's Dilemma games are probably better understood as cases of interdependence. The vampire bats we met earlier regurgitate blood for unlucky roost-mates, not because the hard logic of tit-for-tat demands it, but because vampire bats need to ensure that their buddies will be around on the evenings that they themselves inevitably fail to find food. These need-based systems of help and exchange also seem to have been the norm, rather than the exception, over the course of human evolution and are still common in many hunter-gatherer and other non-industrial societies. They do not replace reciprocal sharing systems, but coexist alongside them. For instance, the Maasai pastoralists of Kenya use two distinct modes of exchanging favours with one another. The *esile* system is reminiscent of the kind of strategy we would expect under a reciprocity-based account of cooperation: individuals help one another, but these are formal debts that need to be repaid. By contrast, within *osotua* ('umbilical cord') relationships (which are formed with a far smaller subset of the entire population), individuals freely give resources to those in need, without any expectation of reciprocation in the future. Such interdependent relationships are common among people living in non-industrial societies, where the availability of food is often sporadic and subject to random shocks. In such circumstances, a friend in need can be the difference between survival or starvation.

Interdependence can also explain why we might even try to *avoid* reciprocity in the context of important relationships. When I was a child, I remember thinking it odd that adults would compete with one another to foot the bill when we ate out at a restaurant. As an adult, the opposite scenario strikes me as far stranger: imagine reminding your friend that you paid for dinner last time, so now it is their turn to cough up. If relationships are built only on reciprocity, then this should strike us as curious: we should avoid the social farce of insisting that it is our turn to pay when it isn't, and we should not feel uncomfortable in pointing out that, on this occasion, it is someone else's turn.

The solution to this puzzle seems to be in the signal that a dogmatic insistence on reciprocity might send. By not counting the pennies in each social interaction, people send an unspoken message to their partners: we are interdependent. Interdependent partners don't need to balance the books in every interaction because interdependence implies that you also have a stake in the welfare of your partner. Insisting that someone buys you a coffee now because you bought them one last week strikes us as petty because it implies that the price of a hot drink exceeds your stake in that person's welfare.

It suggests that we do not place much value on the coffee drinker as a friend.

Nowadays, there are apps that allow friends to invoice one another when they go out for a drink or have one another round for dinner. This technology takes the hassle out of splitting the bill, but might make us feel uncomfortable in doing so. The logic presented here suggests that this unease occurs for a good reason: might we inadvertently be damaging the fabric of our social relationships with such an unyielding approach to give and take?

11
AN EYE FOR AN EYE

> Sooner or later everyone sits down to a banquet of consequences.
>
> Robert Louis Stevenson, 'Old Mortality', 1884

Before the raid, the men gather to feast and prepare for the mission ahead. They plan to ambush the enemy, to kill their men and seize valuable cattle. Amidst the general buzz of excitement at the prospect of victory, some men will be feeling nervous, and scared: battles these days are fought with guns rather than spears, and the risk of serious injury and even death is high.

The Turkana are semi-nomadic pastoralist people, who inhabit desolate territory in northern Kenya. Drought can persist for most of the year and hunger is the norm rather than the exception. To survive, people rely on their farmed livestock and, in extreme scarcity, make incursions into neighbouring territories, to steal cattle to increase the size of their own herd. Partaking in these raids is dangerous: around one in a hundred men will be killed during the fight. Joining a raid is also a form of cooperation because the cattle and other benefits to be gained from intimidating the enemy will be shared by many others, even those who don't fight. Some men inevitably succumb to their fear, abandoning their allies and fleeing back to the safety of the home camp.

These acts of cowardice are generally frowned upon. The anthropologist Sarah Mathew has worked extensively with the Turkana, to understand how social norms of disapproval and punishment can coerce men into fighting rather than fleeing. Among the Turkana, many people (and especially unmarried women) are scathing of deserters. In addition to being lambasted as poor soldiers, cowards are judged as being generally unreliable and undesirable as marriage partners. But it doesn't end with social disapproval: sometimes those who abandon their comrades are also brutally punished for their actions. After a raid, there will be a meeting to discuss how to handle the deserters. If punishment is decided upon, it falls to the miscreant's same-age peers to mete out this rough justice. In Turkana society, the punishment will take the form of being tied to a bush and beaten, before being forced to sacrifice a cow for the group. The aim of this corporal attack is to teach the coward a lesson and to make him think twice before fleeing from a fight in future. As one Turkana man put it, 'The only way for him to change is by beating him ... just to talk to him is not enough.'

*

The reciprocal strategies we explored in the previous chapter are unlikely to encourage people to participate in large-scale costly collective activities, like joining a raid. It's not just because the stakes are particularly high for the men who join these battles. Instead, in interactions involving more than two parties, reciprocity works like a sledgehammer when a precision tool is needed. To understand why, imagine that you have been assigned to work on a group project with two other people. One of your co-workers is a conscientious individual who works hard to make sure that the project is going well. But the other co-worker is a 'free-rider' and prefers to let others do the hard work. If you're a reciprocator, you might decide to slack off yourself, thereby repaying the free-rider's poor efforts. But, in doing so, you penalise your hard-working colleague as a by-product. Reciprocity can support cooperation in pairwise interactions but, in groups, it can cause cooperation to collapse entirely.

To demonstrate this effect, I performed a study at my workplace, using my unsuspecting colleagues as the experimental subjects. The location of my experiment was the departmental tea room, and the kitchen sink in particular. As anyone who has ever shared a communal kitchen can attest, keeping a sink free of dirty items is a formidable challenge. A clean sink is therefore a kind of public good: something that everyone benefits from but which is difficult to sustain. Everyone agrees that it is optimal for all users to clean up their mess, but each individual can be tempted to cheat, by leaving his or her dirty items in the sink. When faced with the evidence of other people's laziness, it can prompt even the most conscientious sink-users to defect.

To run my experiment, I snuck into the department early for a few weeks and manipulated the cleanliness of the sink before my colleagues arrived at work. On some mornings I would do all the washing up, leaving the sink spotless and free of dirty items. On other days, I left a few dirty items in the sink and monitored how many additional items were added during the course of the day. Although anyone who has lived in a student house can probably anticipate the results of this experiment, they offer a powerful proof-of-principle nonetheless. On the 'clean' days, the sink tended to stay clean, whereas the addition of just one unwashed pot initiated a downward spiral of defection. When people saw evidence of other free-riders, they felt entitled to reciprocate, by adding their own dirty plates, cups and spoons to the sink. This pattern, widely replicated in both the lab and the real world, highlights the limits of reciprocity for supporting cooperation in large groups.

These field-based approaches are interesting, but to better understand how people resolve the conflict between personal and collective interest in a more tightly controlled setting, we have to run similar studies in the lab. The 'kitchen sink' scenario can be modelled using an experiment known as the Public Goods Game, which is another kind of social dilemma. It has much in common with the Prisoner's Dilemma game; the main difference being that the Public Goods Game involves several rather than just two players.

In the Public Goods Game, the experimenter gives all the players an endowment at the start of the task and players can then choose how much to invest in a joint account and how much to keep for themselves. Any money invested in the joint account is multiplied by the experimenter and then divided back among all players, regardless of whether they contributed or not. Just like the one-shot Prisoner's Dilemma, 'defecting' is a profitable strategy. Any player who wants to make the most money in this game should invest nothing in the joint project, while free-riding on the investments made by others. The most common outcome in games like this parallels the fate of a shared kitchen sink: many people initially start out by cooperating, but the accumulation of free-riders in the group tends to undermine cooperation in the long run.

Reciprocity therefore seems to have a fairly limited remit for producing cooperation among non-relatives. It fails in group settings and (as we will see in the next chapter) it also performs poorly in scenarios where people don't expect to interact repeatedly, or when the incentives to defect are particularly high. In fact, if reciprocity was the only tool in our box for solving social dilemmas, our cooperative worlds would likely be far smaller, consisting of just a core circle of family and important friends.

Our ability to widen the cooperation net stems from something different: our ability to invent new rules (or institutions) for the games that nature gives us. Institutions are like the icing on a metaphorical cake: layering them on top of a social dilemma changes the appearance and the nature of the interaction. Institutions are a way of changing the rules, allowing us to convert a scenario where everyone's best option is to defect into one where individuals succeed by cooperating.

One of the most important institutions for changing the incentives in social dilemmas is punishment. Adding a threat of punishment to a social dilemma fundamentally changes the incentives to cooperate and defect. In large groups, punishment outperforms reciprocity because it can be targeted: it penalises free-riders without also harming the cooperators.

In the early 2000s, economists Ernst Fehr and Simon Gächter ran a series of laboratory experiments to explore whether humans (in this case, a cohort of Swiss undergraduates) would use punishment to govern themselves in these abstract settings. The students played two versions of the Public Goods Game. The standard game was exactly as I've just described: players could choose to cooperate, by investing in the public good, or to free-ride, by keeping their money for themselves. The punishment game differed only in that, after making their decisions, the students could pay one of their laboratory dollars to impose a $3 'fine' on another person. The results were striking. In the standard condition, contributions to the public good started small and declined steadily over rounds as cooperators realised they were being exploited by the free-riders in the group (and so withheld their own investments in response). But, in the punishment condition, investments started out higher from the beginning and remained high throughout the game. These kinds of studies highlight the difficulty in sustaining cooperation: without the right incentives, cooperation is fragile and easily eroded by the actions of self-interested individuals.

A general conclusion from studies like the one above is that punishment promotes cooperation. It does so by transforming the dilemma, making it profitable to cooperate rather than to defect. Although this conclusion is correct in a broad sense, there are a few important caveats to bear in mind when it comes to thinking about how punishment might feasibly promote cooperation in our daily lives. In chess, there is a proverb, that 'the threat is stronger than the execution'. This saying seems to apply particularly well to punishment, both in economic games and when we think about the workings of our institutionalised penal systems. Although the *threat* of punishment does seem to be an important ingredient in promoting cooperation, punishment that is executed can just as easily destroy cooperation as uphold it.

The 'eye for an eye' approach can cause initially minor disputes to devolve into interminable feuds, with detrimental outcomes for everyone involved. In laboratory games, giving players an option to punish one another frequently provokes retaliation rather than cooperation. Because punishment is costly to administer and receive, it tends to destroy wealth in these settings, leaving all players worse off. In the lab, vendettas destroy earnings but in the real world they can destroy lives. One particularly extreme example concerns an initially trivial quarrel (over leaves and rubbish being blown from one house into another) between two neighbours, that started in a rural Chinese village and ended almost twenty years later, in New York, with one of the neighbours being shot and killed by his rival's son. This inability to de-escalate conflicts makes punishment a dangerous, and potentially explosive, tool for trying to enforce cooperation.

Over the course of history, societies have repeatedly converged on rules and mechanisms for limiting the contexts in which punishment can be used, placing limits on who is allowed to punish whom, what for, and how much. Placing constraints on punishment (and outsourcing it to authorities, like courts of law and prisons) can prevent feuds from erupting, but modern punishment institutions frequently fail to reform offenders and therefore fail to promote cooperation in the broad sense.* Punishment institutions that rehabilitate criminals aim to repair damaged relationships, to

---

* There is a possibility that penal systems promote cooperation not by reforming offenders but by deterring others from committing similar crimes. There is a large literature on 'general deterrence' which I do not delve into here, apart from to note that increasingly severe punishments both historically and in modern times are not associated with reduced crime rates.

compensate victims, and to provide routes by which offenders can re-enter the community. These aims are frequently at odds with the Western penal system, where the emphasis seems to be on retribution rather than reform. Retribution may be psychologically satisfying but the societal benefits are limited at best.

*

A safe conclusion, bearing these caveats in mind, is that the *threat* of punishment can help to encourage cooperation (even if it doesn't necessarily reform cheats). But a moment's attention shows that we have merely replaced one puzzle for another. This is because punishment is a service that is costly to provide. In the experimental setting, punishers pay out of their own pocket to sanction other members of the group. In the real world, punishment can be administered by formal institutions, but these are public goods that must be paid for, via taxes. Individuals that take the law into their own hands to punish cheats incur different costs, that might be paid in the currency of energy, time or the risk that the target might retaliate. Punishment can increase cooperation within groups, but this is a benefit that is shared by all group members, whether they invested to police the free-riders or not. Although punishment is an act of harm, this perspective also highlights it as a potentially prosocial act, where the punisher incurs a personal cost to provide what looks like a group-level benefit. The decision to punish others is therefore a second Public Goods Game layered on top of the initial investment game. For this reason, punishment is sometimes referred to as second-order public good, with people who fail to provide it being labelled second-order free-riders.

Viewing punishment as a form of cooperation, or a second-order public good, highlights the circularity of the argument that 'humans cooperate because they want to avoid being punished'. Punishment probably does encourage cooperation – and might well be an important part of the human story, in explaining how we managed to extend our cooperative networks beyond our immediate kith and kin. But offering 'punishment' as the answer to the question of why we cooperate presents another difficulty: how do we account for the existence of punishment mechanisms that seem to have exactly the same problematic incentives as the cooperation they support?

*

One answer to why people punish, despite the costs in doing so, is that we apparently enjoy doing it. The same brain reward centres that are active during prosocial acts of beneficence (as well as other rewarding activities) also light up when people take the opportunity to punish others. Even children seem to get a kick out of watching bad apples get their comeuppance, as shown in a study using a Punch and Judy set-up, where children paid real money to watch one (naughty) puppet take a beating from another one. In fact, the propensity to punish social cheats is so strong that people are sometimes motivated to intervene on behalf of victims, by punishing cheats in scenarios where we weren't directly involved, a phenomenon known as 'third-party punishment'.

It might feel good to give social cheats a telling-off, but this raises the question of why our brains are designed to enjoy meting out these costly and potentially dangerous acts. For the most part, evolution has designed our psychology such that dangerous or costly activities are perceived as being unpleasant, so that we avoid doing them. Pain is your brain's way of telling you that a part of your body is damaged, so that you might act to prevent further injury and to remedy any harm already done. Hunger is another unpleasant feeling that reminds you that it is time to eat. Conversely, actions that carry long-term benefits are usually experienced as being enjoyable or pleasant. This can prompt us to engage in these activities, even when there might be short-term costs involved in doing so. Sex is a good example. Humans have a lot of it compared to many other species, because we tend to find it enjoyable. But, in the short term at least, sex is actually a rather costly thing to engage in. There are time costs – if you aren't having sex, you can be doing other things instead. Sex also potentially increases exposure to disease and, for many creatures, is a moment when you might be particularly vulnerable to predators. Despite these short-term costs, sex has the potential to lead pretty directly to reproduction. It makes sense that our brains are designed in such a way as to make us enjoy something that carries such an obvious downstream benefit.

Cooperating and punishing others seem to be a bit like sex: activities that are costly in the short term but that nevertheless have the potential to yield benefits in the future. This might explain why our brains are designed to make us enjoy performing these kinds of actions. In fact, we now know that punishers *do* stand to benefit from their investments. One of the first concrete demonstrations of this possibility came from a set of experiments that I ran on a small island off the coast of Australia. Unlike much of the existing work in this field, my experiments were not performed on undergraduates, or even on humans: they were done on fish.

*

On the Great Barrier Reef, you'll find a little fish – the bluestreak cleaner wrasse – with which we have a surprising amount in common. I have studied cleaner-fish behaviour with my colleague, Redouan Bshary, since 2010. Our research is conducted at Lizard Island Research Station, a tiny dot in the South Pacific Ocean, about an hour's flight from Cairns in Far North Queensland. Lizard Island was named by Captain James Cook in 1770 after the large – and slightly intimidating – monitor lizards that roam freely there. Nevertheless, given the names that Cook bequeathed to some of the other places on this difficult part of his voyage (Cape Tribulation, for example, or Weary Bay), I always assumed he must have liked Lizard Island. It's not hard to see why. Pictures taken from the air show a small crescent of land, fringed with white beaches and coral reefs, surrounded by azure sea. It looks idyllic; it did not take much to persuade me that this would be a good place to do field work.

Unfortunately, daily life on a tropical island was far less glamorous than I had imagined. In order to truly understand the behaviour of the cleaner fish, we need to study them in the lab.* Cleaner fish readily adjust to their temporary accommodation, and quickly learn to feed on pulped prawns smeared over the surface of rectangular Plexiglas plates. Rather than sipping piña coladas on the beach, most of my Lizard Island days were therefore spent either mashing prawns, or cooped up in the rooms housing the cleaners' aquaria, fending off mosquitoes while making detailed observations of behaviour.

Cleaner fish are like the hairdressers of coral reefs. They hold small territories, which we call cleaning stations, where they provide a service to their 'clients' – the other fish that live on the reef. The service involves removing ectoparasites and other nasties from the surface of clients' skin, and sometimes even throwing in a relaxing pectoral-fin massage for good measure (which lowers stress in fish just as it does in humans). Regular cleaning is important for the clients' health, and cleaners usually provide a good service to their clients. However, they occasionally also cheat by taking a bite of the client's living tissue, its mucus and scales. Mucus is more nutritious than ectoparasites and also contains UV-absorbing amino acids, which provide the cleaner fish with an ingested form of sunscreen. Although it doesn't sound appetising, mucus is also apparently more delicious than ectoparasites: if you allow a cleaner fish to feed on an anaesthetised client who does not respond to being bitten, the cleaner fish will mostly eat mucus and scales, rather than parasites.

* Don't worry – we returned the cleaner fish back to their reef at the end!

So, there is a conflict of interest between cleaners and clients: clients want the cleaners to remove ectoparasites, while cleaners would prefer to eat mucus and scales. For cooperation to be maintained in this system, this conflict must be resolved. On the face of it, we don't appear to have much in common with these small, reef-dwelling fish, but they face similar social problems to humans living in large-scale societies. In both species, individuals frequently interact with strangers in fleeting encounters, where cooperation is the best mutual outcome but where there is also an omnipresent temptation to defect. It might come as a surprise to learn that, in both species, evolution has independently discovered similar mechanisms for enforcing cooperation. The similarities between cleaner-fish behaviour and our own also highlights a more general point: closeness on the evolutionary tree of life is not always a good predictor for where we will observe behaviours that we think of as being 'human-like'. Instead, taking a broader ecological perspective – asking where the physical or social environment is likely to have favoured certain forms of behaviour – can often be more productive for comparing the behaviour of humans with that of other species.

*

Punishment turns out to be one key way that underwater cooperation is maintained. To understand why, we need to examine the cleaner-fish way of life. On their home territories, cleaner fish live in groups comprising a single male and a harem of females.* Although cleaners typically perform solo inspections of the clients

* Interestingly, all cleaner fish are born as females. Like many fish, they are able to change sex and become male once they reach a certain size. Cleaner fish females turn into males if the current resident male disappears or if they manage to outgrow the resident male on the territory. The fact that a breeding partner can become a competitor further exacerbates the conflict between male and female cleaners in feeding interactions.

that visit the cleaning stations, the male cleaner occasionally works together with one of his females to perform a joint inspection of a large client fish.

If you put yourself in the shoes – or possibly fins – of the client fish for a moment, you might not be particularly enthused about the prospect of being serviced by two cleaners, rather than just one. To see why, consider the incentives faced by each of the cleaners in this scenario. We already know that cleaners prefer to eat mucus rather than removing ectoparasites. If either of the cleaners succumbs to this temptation, and bites the client, the client is likely to swim off in a huff. And, when lunch swims away, both cleaners suffer. This scenario is therefore an underwater Prisoner's Dilemma, which might be expected to result in a race to the bottom with each cleaner trying to gain the benefits of biting the client before the partner beats them to it.

Of course, the real loser in all of this is the poor client! By rights, clients should be very wary of this situation, expecting to receive a terrible service from a pair of cleaners, rather than the five-star treatment they want. But in reality, we see the opposite. Clients that are cleaned by a pair of cleaners receive a much *better* service than those who are cleaned by singletons – and nearly all of the improvement in service quality is provided by the female cleaner fish. What is going on?

Initial observations made while watching the cleaners underwater suggested that males were keeping an eye on proceedings during these joint inspections. If the client jolted (a sign that it had been bitten) or fled from the cleaning station, the male cleaner seemed to blame the female. Following the client's departure, the male would harass the female by chasing her around and trying to tear chunks from her tail with his sharp teeth. These observations looked very much like the male was punishing the female – a curious behaviour given that it is the client, not the male, who is the real victim of the biting female. A male cleaner intervening on behalf of a victim client to punish the female sounds a lot like third-party punishment in humans: a behaviour that has been touted as an altruistic act designed to benefit the punisher's group, and a key ingredient in the rise of our large, ultra-cooperative societies. Is it possible that cleaner fish might be doing something similar? To assess what was going on – and whether males really were punishing females – we needed to explore this behaviour in more detail in the lab.

As I mentioned, cleaners can be trained very easily to feed from Plexiglas plates. In the lab, these plates become our model clients, offering cleaners two different kinds of food: mashed prawn (which they love) and a mush made from commercial fish flakes (which they don't like nearly as much). The fact that cleaners have these preferences is helpful because it means we can create a scenario akin to that which they face when interacting with real clients: the choice between a preferred and non-preferred food. We train the cleaners that if they eat fish flakes, they can continue feeding off the model client but, if they eat a prawn item, the model client 'flees' (this involves me – the experimenter – quickly removing the plate from the aquarium, using a long handle to which it is attached). So, in our lab experiments, eating fish flakes is the equivalent of eating ectoparasites; while eating prawn is the equivalent of cheating, by eating mucus. Remarkably, cleaners learn very quickly – over six trials or fewer – to feed against their preference on model clients. Part of the reason the cleaners learn so readily in the lab is because the rule we teach them has the basic features of the decision they face almost 2,000 times per day on the reef with real client fish: deciding whether to eat preferred food, or not.

In 2010 my colleague, Redouan, caught eight mixed-sex couples of cleaner fish and brought them into the lab, ready for testing on our punishment paradigm. In a rather painstaking process, I spent several weeks that summer using medical tweezers to place minuscule items of flake and prawn on the surface of the plates, before lowering the plates into the aquarium for the cleaner fish to pick them off. To model the scenario where a pair of cleaners inspect a joint client, I allowed the cleaners to forage on the same plate until one of them ate a prawn item (the equivalent of cheating). At this point, I swiftly removed the plate – mimicking the scenario where a client flees from the cleaning station.

In the punishment condition, males and females foraged on the same model client and the male was therefore free to chastise the female if the interaction ended badly. Males responded to female cheating with intense aggression, often chasing the female round the tank. Although males frequently cheated as well, they could more easily get away with it. Females are smaller than, and subordinate to, males and never punished their errant male partners. Male cleaner fish are therefore a bit like hypocritical bullies: they want the females to do as they say and not as they do. The poor females were often quite harassed and intimidated by their male partners, frequently resorting to hiding in their shelter tube. But the male punishment seemed to have the desired effect on female behaviour. After an aggressive reprimand, females tended to behave more cooperatively in subsequent trials: eating more flake items and resisting the temptation to eat the prawn.

But to really test the prediction that male punishment causes the female to change her behaviour, I needed a control scenario: where males and females could feed on the same plate, but where I could also prevent males from punishing the female partner. To do this, I fashioned some transparent plastic barriers that slotted neatly into the cleaners' aquaria. These barriers confined each fish to their own half of the tank, while allowing them to interact with the same model client. This design meant that the male and female could see one another and co-feed on the same model client but, when the interaction ended, the male could not aggressively chase the female. As we expected, in the trials where males were prevented from punishing the females, the females behaved with impunity: they continued to cheat, by eating what they wanted.

Despite apparently intervening on behalf of the client, the male cleaner's motives are far from altruistic. Males clearly have a vested interest in how the female behaves. In addition to the untimely departure of lunch, clients remember when they have had a bad service at a cleaning station, and those who have other options might not come back next time. Male cleaners therefore benefit by forcing their female partners to toe the line (even when they don't do so themselves). In our experiments, males were able to eat more food from the joint client plates when they foraged with cooperative females: in principle, the same is likely to be true in the interactions on the reef with real clients.

*

This cleaner-fish study offers a different perspective for understanding the reasons why humans might punish one another in social dilemmas. One common view is that human punishment is tailored to achieve group-level benefits – to make the communities we live in more cooperative and therefore more successful. However, a simpler alternative is that punishers benefit *personally* from these investments – irrespective of whether punishment is beneficial at a group level. One obvious way that punishers might benefit from their investments is by acquiring a punitive reputation. Punishing others shows that you do not tolerate cheating, and people are more likely to cooperate (or capitulate) when faced with a punitive opponent. Punishment might also send a different sort of signal to onlookers: that the punisher is a fair-minded individual who is prepared to incur a personal cost to enforce cooperation (think Robin Hood, for example). Again, laboratory experiments show that punishers (people who spend some of their experimental money to impose a fine on cheating or free-riding players) are more likely to be trusted in a cooperation game, and are also more likely to be rewarded by others for their heroic deeds. In sum, these studies cast doubt on the idea that investments in punishment are actions that can only be understood by appeals to group-level benefits and show that, in many cases, punishers might be pursuing their own self-interest after all.

Reputation doesn't just help to explain why we punish others: it accounts for lots of the strange things that humans do. In the previous chapter, we met Stephen and Sarah – the two contestants in the finale of the television show, *Golden Balls*. Since then, we've seen that humans often cooperate to sustain important relationships and to avoid the threat of punishment from others. But neither the logic of repeated interactions nor the threat of punishment can explain why Stephen chose to split the money with Sarah. The *Golden Balls* finale is a definitive one-shot interaction: Stephen and Sarah are two strangers who make a single decision to split or steal and who will then go their separate ways in life. There is no threat of punishment for choosing 'Steal', and the shadow of the future does not hang over this particular relationship.

But there's another reason why it might be a good idea to cooperate, even in an encounter that appears to have little probability of being repeated: your reputation. Part of the reason this particular episode makes for such compelling, if uncomfortable, viewing is that most people are somewhat more concerned with their reputation than Sarah appears to be. You might barely raise an eyebrow if you learned that Sarah had made this decision privately, in a situation where no one, perhaps not even poor old Stephen, would ever find out. But Sarah's decision to shaft someone in public – on television, no less – where other people can see her actions and judge her accordingly, feels far more surprising. In contrast to this flagrant act of self-interest, most people usually behave as though they do care about what others think of them, presenting themselves in a manner that showcases their finer qualities and conceals their flaws. Caring about what others think of us is a part of our psychology that has been honed by selection, but why?

12
PEACOCKING

> The sight of a feather in a peacock's tail, whenever I gaze at it, makes me sick!
>
> Charles Darwin, 1860

If you're like me, you probably check a seller's feedback score on eBay before you decide whether to buy from them or not. Choosing someone with a good reputation score is reassuring because we assume that the seller won't risk their reputation just to make a quick buck by swindling us. In its basest form, reputation is information about another party, allowing us to infer how they might behave in the future, based on what they've done in the past. We judge other people based on their reputation – and we attempt to manage our own image by cooperating more when others might see it, and less when there is a smaller prospect of being observed. Curating a valuable reputation is therefore an investment, but one which pays dividends in the long run. On eBay, a good reputation pays off in real monetary terms: the sellers with the highest feedback scores can demand a higher price for the same product compared to sellers whose reputations aren't quite as shiny.

As you might expect, people do typically behave as if they are aware that their reputation is valuable: countless experiments have shown that people are more likely to behave generously when their acts will be made known to others, compared to when these good deeds go unseen. Savvy policymakers can use these insights to nudge citizens into performing socially desirable actions at low cost. In 2013, a team of US researchers teamed up with a Californian utility company to explore whether reputational incentives could encourage participation in a programme that would allow the utility company to install devices that curbed energy supply during periods of high demand. The utility company had previously offered a cash incentive of $25 to each household that signed up to the demand-reduction programme – an incentive that the company was convinced would be the simplest and most effective way to encourage participation.

But the researchers had different ideas. They suggested placing sign-up sheets in prominent locations in the apartment blocks, allowing people to advertise their participation in the scheme and to monitor who among their neighbours was also signed up. To the utility company's surprise, the simple trick of allowing homeowners to signal their good deed to others was more than seven times as effective as the cash incentive. Removing the feeling that others might see and judge your behaviour can have the opposite effect, making socially desirable actions less likely – and can explain why well-meaning policies sometimes backfire. In Switzerland, attempts to increase voter turnout by introducing postal votes had a limited effect, perhaps because citizens were unable to accrue the reputation benefits of participating in democracy when they no longer needed to go to the polling station to do it.

*

We are not the only image-conscious species on the planet. Once again, the cleaner fish appears to be doing something remarkably similar to us, by trying to present itself in a positive light to others. Remember that there is a conflict of interest between a cleaner fish and its client: the client wants the cleaner to remove the parasites, while the cleaner fish would prefer to eat mucus and scales. Unlike humans, cleaners and clients can't sit down and talk about it and clients cannot leave feedback for bad service. Nevertheless, the tension between the two parties in this system is resolved in strikingly similar ways.

One of the ways that clients can keep a cleaner fish honest is by voting with their fins. Some clients have access to several cleaning stations within their home range and, as such, they do not need to put up with substandard service. These 'fussy' clients behave like prima donnas: they demand immediate attention and refuse to wait for a cleaning service. If they don't like what they see at a cleaning station (for example, if they see an altercation between a cleaner fish and a current client) then they swim away and look for a better service elsewhere.

Divas demand better service – and they get it: whereas the hoi polloi have to wait to have their parasites removed, cleaners frequently promote the fussy clients to the head of the queue when they arrive at a cleaning station. Even more astonishingly, work led by Ana Pinto during her PhD found that cleaners take extra care with their current client when being watched by a fussy client nearby. This demonstrates a rudimentary concern for reputation; something that is extremely rare in nature. Even humans struggle with reputation management until we hit middle childhood, and there is scant evidence that any of the other great apes know or care about what others think of them.

The fact that cleaner fish do this, while other species do not, does not imply that cleaners are somehow cleverer than chimpanzees or human children. Cleaners are unlikely to use the same cognitive strategies to manage their reputations that humans use, and that children develop as they progress through childhood. For humans, reputation management involves taking the perspective of another person, and also inferring how their beliefs and impressions of us might be altered under various scenarios. These scenarios don't even have to happen for us to reason in this way: we can imagine what people might think of us if we were to cheat on our taxes or win a Nobel Prize, as well as scenarios where our hypothetical behaviour goes undetected. These daydreams feel effortless to us, but drawing such inferences is computationally taxing – and cannot be done by any non-human species. Cleaner fish don't solve reputation-management tasks by thinking in this way; rather their skills are more likely to be based on a much simpler form of associative learning: over time they learn that some kinds of clients tend to swim away if they don't get a good service, or if they see the cleaner fish bite another client. To put it another way, a cheating cleaner fish might predict *that* a bystander client will swim away but, unlike humans, they don't understand *why*.

We can also draw a strong parallel here with the findings concerning the occurrence of teaching and the kind of cognition that supports such behaviour. Teaching *can* involve the cognitive superpower of knowing what the pupil knows, but it doesn't have to. Creatures like ants and babblers and meerkats can act as teachers, despite (as far as we know) not being able to think in this way. Reputation management is similar: for humans, it seems to rely on the ability to infer the beliefs and impressions held by others, but it doesn't have to (and in cleaner fish, it doesn't). This seems to be another case where humans and other species seem to reach the same behavioural destination, but via a different cognitive journey.

*

Without systems to track and monitor the reputations of others, it is unlikely that the intricate systems of mutual trade that characterise all human societies would ever have emerged. Like cooperating in a Prisoner's Dilemma, trading in a marketplace is a risky move that only pays off if the partner is a cooperator, rather than a defector. To make the bet look worthwhile, individuals must trust one another. Most trading interactions do not involve the simultaneous exchange of resources, where each partner can give with one hand, while receiving with the other. Instead, trades tend to involve asynchronous moves, meaning that whoever makes the opening gambit is vulnerable to exploitation.

In a restaurant, you do not pay the waiter each time he delivers a plate of food. Rather, the food is provided upfront, with the restaurant 'trusting' that the diner will pay for the meal at the end. In a fast-food joint, things work the other way: the diner pays first, trusting that the food will then be provided. Even if we could figure out ways to perfectly and synchronously exchange resources in every interaction, this would not solve the risk problem since there is commonly an information asymmetry between traders, meaning that individuals can palm one another off with substandard products. These are not just theoretical abstractions: the 2013 horsemeat scandal in the UK, in which a number of supermarkets were discovered to be selling horsemeat products labelled as beef, indicates the extent to which sellers (in this case) can conceal information from buyers, and the resulting loss in trust that can ensue when such transgressions are discovered.*

A lack of trust can hinder mutually profitable trade. One solution is to outsource enforcement to higher authorities; a common solution in modern, industrialised societies, where various bodies (police, courts of law and so on) exist to apprehend and penalise those who violate the social contract. Although these authorities are important and worthy of discussion in their own right, we will leave them to one side for the moment because we know that they cannot be the only explanation for the emergence of trust among trading partners. For one thing, the creation of these higher-order public goods is itself a form of cooperation (the police are paid for by the

* Tesco, one of the main supermarkets involved, was estimated to have had £300 million wiped off its share value in the wake of this scandal.

contributions of taxpayers, after all). To invoke their existence as the key pillar supporting cooperation raises the question of how we solved the problem of supplying these institutions in the first place. What's more, authorities are not always needed to help us resolve social dilemmas: cleaner fish manage without them, and so, too, do humans in many marketplaces.

Consider how trust is maintained on the Dark Web, the online marketplace for illegal goods: narcotics, weapons and so on. A person involved in an online drug deal can hardly complain to the police if a trading partner swindles them and, even if they could, there is little the police could do about it. Interactions in these cryptomarkets are carried out in total anonymity, with traders using encryption software that deletes any data that could be used to identify them or reveal their location. In a marketplace where interactions take place among faceless criminals and where there is no authority to uphold the law, we might expect trust (and trading) to be virtually non-existent. But the existence of reputation systems – involving stars and written feedback, just like other online platforms – allows these markets to flourish. In 2017, for example, a month's worth of sales data scraped from one Dark Web sales platform suggested it would generate approximately $390 million in that year.

Prison gangs seem to serve a similar function – but this time serving to maintain a reputation of a collective rather than an individual. The scholar David Skarbek has argued that although gangs typically form along ethnic lines, they are not used to facilitate inter-ethnic violence: homicide rates in prisons have fallen at the same time as the prevalence of prison gangs has rocketed. The mission of a prison gang seems to be different: to maintain the gang's reputation as a worthwhile trading organisation. This, in turn, fosters cooperation and trust between different gangs, which smooths the trade of drugs and other contraband both in prison and on the outside. Gang members carefully police the actions of their affiliates – any member who cheats or steals or generally tarnishes the gang's reputation is placed on a 'bad news list' and is to be attacked on sight. This policing from within prevents inter-gang feuds from erupting and ensures that trade can continue unhindered.

*

Like punishment, formal reputation systems are also types of institutions: human-devised rules that transform scenarios where trust might ordinarily be low into ones where individuals are mutually reassured that their partners are incentivised to cooperate. Without these systems, platforms like Airbnb, Uber and Lyft would be unlikely to exist. But reputation systems are not a modern invention. In fact, in medieval Europe, institutional mechanisms to keep track of others helped to oil the wheels of merchant trading in foreign territories.

Traders in the eleventh century faced a dilemma when it came to selling their goods overseas: they could personally travel with their wares and sell them on the foreign market themselves, or they could entrust the task to a foreign agent, who would sell the goods on their behalf. The latter was a more efficient option – but carried the attendant problem of trust: how could a trader be sure that the foreign agent wouldn't just take the goods and clear off with them?

The solution came in the form of merchant guilds, such as the Maghribi traders – a club that only admitted the most trustworthy members of society. By choosing to do business with a member of the Maghribi guild, a trader could be sure that his partner was committed to doing honest business. A Maghribi trader would face the much larger cost of being excluded from the guild if he didn't toe the line. People intuitively trust the drivers of London's famous black cabs for the same reason. Cabbies spend two years of their life studying for the devilish test ('The Knowledge') that determines their grasp of the various As to Bs in the city and permits them to become members of the 'Worshipful Company of Hackney

Carriage Drivers'. Attaining membership of this exclusive club acts as a commitment device: the risk of expulsion acts as a strong disincentive to cheat by taking passengers on the 'scenic route' to their destination. Taking things to the next logical step, the rather insidious Social Credit System being developed in China has the same aim: to give every person and organisation a reputation score which can be used to decide which school your children attend, whether and where you can travel, and who you get matched with on dating websites. According to an official document, the system will be regarded a success if it allows 'the trustworthy to roam everywhere under heaven while making it hard for the discredited to take a single step'.

*

Cooperative individuals can reveal their credentials by accruing a good reputation – but signalling systems like these are vulnerable to dishonest brokers. On the reef, cleaner fish have to contend with another species that looks exactly like them: the sabre-toothed blenny – a fish that sounds far more terrifying than it looks. I'm embarrassed to admit that, even as a seasoned cleaner-fish researcher, I was once fooled by one of these mimics for several days. I had been habituating my newly-caught cleaner fish to life in the lab, a process where we tried to help the cleaners become accustomed to their temporary lifestyle. One evening, I mentioned to Redouan that one of the fish was acting in a strange manner, preferring to hide away in the shelter tube for most of the day rather than swimming around the tank like the other cleaners did. Eventually, Redouan came to inspect my strange fish and burst into laughter: I had been fretting over a blenny! Its odd behaviour was immediately obvious: most blenny species are ambush attackers that hide in small shelters on the reef and wait for an unsuspecting victim to swim by, before darting out and tearing off a chunk of flesh with their sharp teeth. The blennies make use of the fact that they look like a species that clients trust – the cleaner fish – which allows them to approach and strike at close quarters.

Sometimes cleaner fish themselves behave dishonestly, tricking clients into trusting them. We saw earlier that cleaners obtain nutritional benefits and also a form of sunscreen when they ingest the mucus of their clients. But not all clients are equal in this respect: some have richer mucus than others. When cleaners are very food-stressed, they occasionally fool one of these highly valuable clients into accepting an interaction, by giving another fish nearby a relaxing fin massage. The signal being sent seems to be 'Look – I'm giving great service to my clients. You can trust me.' But this is a case of false advertising: as soon as the client with the tasty mucus approaches, then the cleaner fish swipes a chunk of this valuable slime.

When there are benefits to be gained from signalling, then dishonesty is always a possibility. Remember how easily Sarah convinced Stephen to trust her, to his eventual cost. The old adage that 'talk is cheap' turns out to contain a deeper biological truth: signals that are easy to fake are often unreliable. Costly signals are inherently more credible because they are either impossible or unprofitable to fake. Take the wealthy Indian businessman, Mr Datta Phuge,* who became famous for wearing a $250,000 shirt made entirely from gold. As a functional item of clothing, the shirt performed abysmally: it weighed more than a newborn baby, it couldn't be washed, and Mr Phuge had to enlist a team of bodyguards for protection whenever he wore it. But, as a signal of wealth, the shirt was highly effective: who else but the super-rich would dream of purchasing and wearing such an item?

*

* Mr Phuge was murdered over a dispute involving money in 2016.

Flamboyant displays are often costly signals of an otherwise hidden quality. In the natural world, the species that most readily comes to mind as an insufferable show-off is the peacock. Spanning one and a half metres and weighing around 300 grams, the male's tail seems to serve no purpose other than to convince admiring females of his desirability. But size isn't everything: we all know it's not what you've got but what you do with it that counts. A peacock's large tail won't necessarily bring all the hens to the yard. If only it were that easy! To woo the females, an ardent male must perform an energetic display, vibrating his tail feathers at just the right frequency for several minutes, and aligning himself so that the sun glints off the iridescent eyespots on his tail feathers at just the right angle to lull the females into an awed stupor. The sheer extravagance of the display bakes the honesty into it: it is only worth going to such extreme measures if you really are a high-quality individual. Lesser cocks need not apply. The peacock's tail feather may have sickened Darwin (because he couldn't reconcile the extravagance of the seemingly wasteful display with his theory of evolution by natural selection), but it provides the peahen with crucial information about a potential mating partner. Females benefit by being circumspect about whom they mate with: choosing the showiest males means that they will give birth to higher-quality offspring.

Costly signals are not always tail-shaped: the stereotypical stotting a gazelle performs when chased by a predator serves as an honest signal of its athletic prowess and, therefore, its ability to evade capture. The energetic begging displays of nestlings are also subject to similar dynamics, with the costs of investing in these vigorous displays placing a theoretical limit on the incentive for chicks to misrepresent their true hunger to the parents. The costs can be direct: it is not worth expending more calories on a display than you will obtain from receiving a beakful of food. Costs can also be indirect, however, because a morsel of food received by one chick is a morsel that won't be placed in the mouth of a sibling, either current or future. Well-fed chicks might 'prefer' that their siblings receive the food item, rather than vigorously (and dishonestly) begging to secure it for themselves. As you might expect, the potential for dishonest signalling increases when chicks are less related to their brood-mates or future siblings (and therefore less concerned with their siblings' survival).

The bowerbirds of Australia go so far as to construct miniature show homes to attract females, as well as destroying any particularly lovely bowers built by the neighbours (to kill off the competition). Males keep their bowers meticulously tidy, adorning them with items that females like – berries, shells, even pieces of broken glass; the decorations arranged in prominent locations to be sure to catch the female's eye. Like the impractical gold shirt, the bowerbird's show home is not intended to be used as accommodation; neither the male nor the female will live in it. Instead, this extravagant palace is nothing more than a costly signal – a way for the male to catch a female's eye and (hopefully) increase his reproductive success.

Many of the signals we send to bolster our own reputations might also adhere to these principles, and much of my research on humans has been aimed at understanding our apparent obsession with reputation – and how this might affect our tendency to play nice in social interactions. I want to stress again that this should not be taken to imply that humans always or even routinely calculate the likely reputation benefits of performing cooperative actions any more than, say, a mother calculates the survival benefits of feeding a crying child before raising it to her breast. We can accept the premise that people behave *as if* they show concern for reputation without also insisting that all good deeds are performed in a calculated manner to achieve this goal.

*

access to larger social networks and more support within them – benefits that are likely to be as relevant to women as they are to men. Among the Ache of Paraguay, those who shared a greater proportion of their food with others were more likely to receive help in periods when they were sick or otherwise unable to obtain food. Status benefits are also available to all hunters, regardless of their skills: a frequently observed pattern is that the best reputations are earned by the most generous individuals, rather than the most productive.

In the Martu of Western Australia, it is the women that embark on cooperative hunting expeditions to capture live prey, such as monitor lizards (and, more recently, feral cats). When the hunters return, the camp-mates gather round a communal fire to share their evening meal. As each item of cooked meat is pulled from the flames, it is handed to the woman who caught it – but, rather than keeping it for herself, there ensues a ritualised process of exchange, with the hunters and others in the group handing off and receiving meat from others, until all the meat has been distributed more or less equally around the fire, regardless of who caught it. The common – and intended – outcome of this redistribution is that the most successful hunters receive slightly less than anyone else around the fire. Once again it is the most generous, rather than the most skilful, hunters who others choose to associate with. By virtue of their generosity, these hunters invest in social capital, building a support network that they can leverage in times of need.

It makes sense that people pay attention not just to signals of expertise, but to signals of generosity. After all, the best social partners are not just those who are able to help you – but those who are also willing. Some of my own research (with my colleague, Pat Barclay) supports the idea that reputation is built on generosity rather than absolute wealth or prowess. Our study used the scenario just described in the Martu as a starting point: whether to interact with someone who is an excellent hunter, but seldom shares the spoils; or with someone less skilled, but who shares whatever they get. We turned this into a laboratory game that allowed us to show that people really do prioritise a partner's *willingness* to help over their *ability* to do so. In our experiment, people had to choose one of two partners to interact with: either a poor-but-fair person, who shared half of any winnings with the participant; or a rich-but-stingy person, who won a larger prize, but shared less of it with the participant. We found that people preferred to interact with the nice guys, even when choosing the richer (but stingier) partner meant that participants would probably earn more money in the task. This supports the results from the field studies described above, by suggesting that people place more emphasis on a partner's willingness than on their ability to share. This preference is understandable in a world where the ability to share might be quite variable. Even the best hunters in hunter-gatherer societies have high failure rates; and wealth can come and go. What matters more than what you have is whether you share when you can.

*

I have also been particularly interested in the possibility that cooperative signals might be used to attract members of the opposite sex. We saw just a moment ago that men can gain status benefits from signalling their cooperative natures – but do women really care about this? The available data suggests that they do. Since the late 1980s, scientists interested in the potential for sex differences in mating preferences have accumulated an impressive arsenal of data from surveys, from speed-daters, from personal ads and even from mail-order brides, all aimed at testing whether men and women prioritise different attributes when seeking a sexual partner. The findings of these endeavours have been remarkably consistent, both across time and around the world: men and women typically place different emphasis on the attributes they prioritise in opposite-sex partners. There are some traits that both men and women agree are desirable (e.g. kindness and loyalty), but men typically place an additional premium on indicators of fertility (which correlates with youth) and attractiveness. Women, on the other hand, prioritise attributes that signal status or resources, preferring men who are slightly older and who are either wealthier or have greater earning potential. Men are from Mars and all that.

Using these well-established insights as a springboard, I began a project (with an economics professor, Sarah Smith) to study the signalling benefits of donating to charity. Charitable giving falls squarely under the banner of 'puzzling prosocial behaviour'. By definition, giving to charity involves a cost, and one which you are definitely not expecting to be repaid by the beneficiary. Reciprocity can't be at work but, like hunting, this is an arena where prosocial preferences are favoured at least in part because of the reputational benefits that helping confers on donors. We wanted to push this logic one step further, asking whether people might compete to acquire these reputation benefits and whether people might therefore engage in generosity tournaments, by trying to out-do one another in the giving stakes.

To understand how signals might be used to compete with rivals, it might be helpful to think about peacocks again. The peacock's magnificent tail and his titillating shimmy-shimmy dance seem to capture the females' attention but, you might wonder, why does the tail have to be *so* big? Couldn't a tail half the size, still costly but not quite as ridiculous, do the job? Does he really need to bother with all the strutting and elaborate feather-rustling, rather than simply revealing his tail to the female ('Look at my honest signal!'), and then hiding it again? The answer lies in the economic forces of supply and demand. A single peacock is, in principle, capable of siring the offspring of several peahens. This means that a relatively small fraction of the males in a population can theoretically gain a disproportionate share of reproduction, with many others getting no action at all. A peacock's reproductive success depends on being chosen by the hen, and females care about – you guessed it – the extravagance of the male's display. If peahens tend to choose the males who have the largest tails and perform the showiest displays, then selection will tend to favour genes that (when expressed in males) amplify these traits until the point at which the cost to the male of carrying such a large appendage and waving it about becomes prohibitive. Under competitive regimes like this, exaggerated signals can flourish because of the effects they have on the bearer's reproductive success.

*

It turns out that similar selective forces might also be at work in humans. Whether we are talking about a sexy tail or a charitable donation, the underlying logic is similar: if women tend to prefer men who signal that they have resources or are willing to share them, and this leads to differential reproductive success for men with a generous reputation, then competition in this mating market could also result in bidding wars, where generous actions, like peacocks' displays, escalate over time.

To test this idea, we used online fundraising platforms, the kind that are used by people who perform a personal challenge in order to raise money for charity (mostly by badgering their friends, family and other acquaintances for donations). Each fundraiser creates and hosts a personal fundraising webpage, where they tell people what they are doing (a marathon, or a triathlon, for example) as well as the charity they are supporting and the amount they hope to raise. Crucially, the fundraising pages show all the donations that the fundraiser has already received, as well as the amount that each donor gave. Fundraising webpages are therefore an excellent place to look for evidence of competitive helping since potential donors can see who has donated previously and how much they gave. Donors use this information to decide how much they want to give themselves.

We expected that men would be most likely to enter into generosity tournaments in response to seeing a large donation left by a previous male donor on the fundraising page. But we thought the signalling might be a bit more nuanced than this. The most obvious audience of any generous donation on a fundraising page is the fundraiser themselves – and so we predicted that men would be sensitive to whether they were donating on the page of a female or a male fundraiser, being most likely to escalate their own contribution when the fundraiser was a particularly attractive member of the opposite sex.

To test our idea, we used data from over 2,500 online fundraising pages set up by people who had run the 2014 London Marathon. On each of these pages, the fundraiser had uploaded a single photo of themselves, which we had independently rated for attractiveness. We then looked at the donations that arrived on the fundraiser's page. The first thing to establish was whether new donors paid any attention to the size of the previous donations on the page, which they did. Even if the average donation on a page was around £20, a large donation of £50 or more prompted an upswing in the subsequent offerings of about £10, as new donors arriving on the page anchored themselves to the new norm. Small donations had the opposite effect, reducing the size of subsequent donations on the page.

So far so good, but what about competitive helping? Remember that we predicted that men would be most likely to get into these giving competitions – and that this would be most pronounced when they were donating on the pages of attractive female fundraisers. To test this, we looked at what happened after a large donation (defined as being at least £50, or at least twice the average donation amount on that page) arrived on a fundraising page.

Our expectation was that men would be responsive to these showy signals, but only when the large donation had been made by another man and when the donations were made on the pages of attractive female fundraisers. That's exactly what we found: in the 'peacocking' situation, men's donations almost quadrupled in size.* To reiterate, these results should not be taken to imply that men are making these decisions in a conscious, calculating manner, nor that all charitable giving is only attributable to signalling benefits, but it does give us some insight into how concern for reputation could have shaped helpful tendencies in our species, and why.

Despite the evidence that people act to preserve or enhance their reputations, you might be reading this with one half-raised eyebrow. Sure, people care what others think of them – no one wants to come across as being a total jerk. But a concern for reputation cannot be the only reason we do nice things, like sharing food or giving to charity. People often perform good deeds when there is no chance whatsoever that others will find out, and we sometimes even try to hide our generous actions from others. For instance, on the fundraising pages I just described, there is often an option to donate anonymously, and around 12% of people who donate online choose to hide their donation from others. From all that we've learned so far, it would seem that this is a strange kind of own goal: performing a costly, prosocial behaviour and then opting out of the reputation benefits that would otherwise come your way. Although it might appear counter-intuitive, I think that the fact that people often conceal their good deeds from others can still be understood in light of a concern for reputation. I'll explain why in the next chapter.

* What about women, though? In our study, we looked for evidence of competitive giving among women as well, but didn't find it. This doesn't mean that status and reputation are unimportant to women (remember the Martu hunters, for example) but rather that women use different strategies to achieve these goals.

13
THE REPUTATION TIGHTROPE

> We are evidently evolved to deny that we have evolved to be genetically self-serving.
>
> Richard D. Alexander, 'The Challenge of Human Social Behaviour', 2006

In 2014, the word 'humblebrag' was added to the Oxford online dictionary, along with the following definition: 'An ostensibly modest or self-deprecating statement whose actual purpose is to draw attention to something of which one is proud.' In the wild, humblebrags often present as false complaints ('I've lost so much weight I have nothing to wear!' or 'So stressed: I applied to six jobs and got all of them!') or as a boast cloaked in humility ('I can't believe my book became a bestseller!'). Another kind of humblebrag is the one where we tell people what a wonderful person we are, without coming right out and saying it. For example, take a look at the following (genuine) tweet: 'I just did something very selfless. But more importantly, it was genuine & I know it means a lot to the person in the long run #SoWorthIt.'

If such a statement prompts a wry smirk, you're not alone. Brazenly announcing one's virtuosity typically invites scepticism rather than adulation. Even children as young as eight years old take such self-serving claims with a pinch of salt, attaching a higher moral value to individuals who perform good deeds in private rather than bragging about them in public. Experiments with adults find that the perception of someone's generosity is downgraded when they broadcast their good deeds on websites like Facebook. Oscar Wilde put it best with his assertion that the 'nicest feeling in the world is to do a good deed anonymously – and have somebody find out'.

There is an unscientific explanation for all of this, but which nevertheless makes it easy to understand: humans are intuitive bullshit detectors. We don't take actions at face value. Instead, we attempt to look under the hood, to impute thoughts, emotions, beliefs and desires to the person performing the behaviour. The ability to manage our own reputation is built upon our unique potential to see the world from another person's perspective, and to ask how they might update their beliefs about you in light of the behaviour they observe (or hear about from others). And we use these same cognitive skills to try and understand why other people behave as they do.

Evidence that this relies on fairly sophisticated socio-cognitive abilities comes from studies of young children. Human children are not born with the ability to make inferences about the mental states of other individuals or to feel embarrassment or shame about their actions. Instead, the ability to take the perspective of another individual is something that emerges during development. Before the age of five (or thereabouts), children don't really know or care what other people think of them and don't attempt to curate their own reputation at all. It is only when they are around eight years old that children start to understand fully how their actions make them look, and to interpret other people's prosocial behaviour in terms of self-serving motives. By contrast, chimpanzees don't attempt to strategically manage their own reputation at all, never mind second-guessing the motives behind benign actions performed by others.

*

As we've seen already, humans and cleaner fish manage reputation in very different ways. It is extremely unlikely that cleaner fish have anything even remotely resembling this kind of cognition. But then again, they probably don't need it. The rules that govern their interactions with clients can be learned in a trial-and-error fashion and, with thousands of interactions per day, cleaners have plenty of learning opportunities. To put it another way, a cleaner fish doesn't have to read the mind of a client to learn that cheating doesn't pay: they can simply put two and two together and learn that eating mucus makes some fish swim away. These cognitive differences start to shed light on some of the differences in reputation systems in a species like us, that can mind-read, and a species like the cleaner fish, that can't.

For instance, because we understand that good deeds can heap benefits onto the beneficent individual, we try to infer whether these acts were performed in pursuit of these reputation or prestige benefits (in which case we frequently withhold them). Because cooperators can take the moral high ground, they are sometimes treated with antipathy, even when the motives underlying their actions are not in question. For instance, consider the relentless ridicule, derogation and even sarcastic death threats* that vegans get from meat-eaters. Despite the well-known joke,† it is highly unlikely that vegans opt for this diet on the basis of reputation concerns. Reducing meat consumption is morally worthy not just because it reduces animal suffering but also because cutting meat out of the diet is thought to be the single biggest lifestyle change that can help to reduce personal carbon emissions. Why do people enjoy denigrating vegans, when we should really be applauding them?

---

* In 2018, the former editor of the Waitrose *Food* magazine, William Sitwell, resigned for sending an internal email in which he joked that he'd like to print a feature on 'killing vegans, one by one'.

† Q: How do you know if someone is a vegan? A: Don't worry: they'll tell you.

Vegans might take heart in the knowledge that 'do-gooder derogation' can also be elicited in highly abstract laboratory scenarios – regardless of what people do or don't eat. For instance, in a Public Goods Game, where people could cooperate by contributing their money to a collective pot, or defect by keeping their money for themselves, group members frequently reported disliking the person who contributed the *most* to the collective pot, adding that they would like to kick them out of the group if possible. When asked to explain this resentment, people said things like 'No one else is doing what he does. He makes us all look bad' and 'This would be OK if someone else in the group was being like this, but no one is so it's wrong.' When given the chance, some people take things one step further by paying to punish the most cooperative members of their group. This 'antisocial punishment' was originally dismissed as an experimental anomaly but is routinely observed (to varying degrees) in these kinds of experimental settings all around the world, and is thought to be a tool that punishers use to elevate their own rank in the game, relative to others. It's still a status game, however we play it.

*

With this in mind, things like anonymous giving start to make more sense: a donation that far outshines that of everybody else on the page might draw the wrong kind of attention. In 2014, I ran a study using donations to online fundraising pages, where I found that the tendency to give anonymously was not distributed evenly among all of the donors. Instead, as you might expect, people were more likely to donate anonymously when they were making very small donations but also when they were making excessively large donations, relative to what others had already given on that page. The desire to avoid being perceived as a braggart can also deter people from telling others that they've donated to charity, even though posting these donations to social media is known to generate an influx of donations from that donor's social network. Back in 2010, the fundraising platform JustGiving had worked out that every Facebook 'like' on a fundraising page shared to the platform was worth around £5 in additional donations. Inspired by these stats, the team tried to nudge donors into sharing their friend's fundraising page on Facebook immediately after they had donated. But people seemed to find sharing their beneficence a bit cringeworthy and were reluctant to pat themselves on the back in such a public forum.

To try to encourage donors to share their donations on social media, the JustGiving team ran an experiment where they manipulated the message they showed to the donors. One of the least effective prompts was one inviting the donor to publicly congratulate themselves, 'You're an amazing person. Share your donation!' 'Think your friends might care about this too?' was similarly fruitless. The one that worked best? 'Help your friend raise even more money by sharing their page!' This message works because it gives people the permission to advertise their good deed, while maintaining their own sense that they are doing it for the right reasons: to help a friend rather than to show off. This simple change in wording increased the tendency to share to Facebook by 28%, leading to an estimated £3 million increase in charitable donations over a single year. Not a bad return on investment.

So, we see that virtuous behaviour is not a silver bullet to gaining prestige and status benefits. People frequently try to infer the motives underpinning good deeds, and excessively generous acts can be perceived as competitive rather than altruistic. This 'tainted altruism' effect can produce some highly suboptimal outcomes, particularly for people or companies that work in the for-profit, for-good sector. Take Pallotta Teamworks, a fundraising company that described itself as being motivated by 'asking people to do the most they can do instead of the least'. Established in 1982 by Dan

Pallotta, the company pioneered an innovative approach to raising money for charitable causes. Rather than allowing fundraisers to raise any amount they could, Pallotta challenged people to complete unique multi-day events, like the Breast Cancer Three-Day Walk, and to commit to raising four-figure sums in doing so. Over the course of nine years, the approach netted a staggering $305 million for the charities involved. However, there was a major PR problem: Pallotta Teamworks was a for-profit company, not a charitable organisation itself. In 2002, as news of Dan Pallotta's six-figure salary and the sums of money the company was making started to emerge, the public outcry forced the charities to sever links with Pallotta Teamworks, and the company folded. Ironically, the charities also suffered vastly reduced revenues as a consequence. JustGiving has encountered similar ire from the public for being a for-profit, despite helping charities to raise hundreds of millions of pounds for good causes.

This is not intended to be a defence of companies like Pallotta Teamworks or JustGiving. I use these examples to illustrate how quirks of our psychology can prompt a form of moral hypocrisy. We say we like people who do good things, but then we make fun of them or try to exclude them from the group. We say we think it is good to raise money for charity or protect the environment, but we rail against companies that try to achieve these aims if they also derive a profit in doing so. Our difficulty in reconciling the fact that something can be both for profit and for good at the same time frequently prompts us to choose outcomes or people or companies that deliver no benefit whatsoever to good causes, rather than those that take a slice of the benefits they generate. The knowledge that our gut reactions to these kinds of scenarios can sometimes lead to objectively worse outcomes might prompt us to pause and evaluate our reactions to good deeds, before reflexively damning the morally superior individuals and organisations that benefit from investing in prosocial ventures.

*

One of my favourite examples of how advertising good deeds can completely backfire is a short piece written by the anthropologist Richard Lee, 'Eating Christmas in the Kalahari'. Lee had been working with the Ju/'hoansi bushmen of the Kalahari, studying their traditional methods of hunting and gathering. During this time, Lee had gained a reputation for being tight-fisted, in part because his commitment to data collection meant he did not share his own supply of food and other goods with the people whose mode of subsistence he was studying. To make amends, at Christmas, Lee decides to supply the meat for a feast. He buys the biggest ox he can find, certain that – in this resource-scarce environment – his offering will be greatly appreciated. But, rather than being revered and thanked, the bushmen haughtily dismiss the gift, telling him the ox is a bag of bones and too scrawny to feed the entire camp. He hears complaints that people will surely go to bed hungry and sad if that is all that is on offer and one man warns him that the great ox is so insufficient that it is sure to provoke fights over the scraps of meat.

Lee spends a few days frantically searching for a second beast to add to the first, to no avail. It is not until Christmas Day itself that he realises the extent of his faux pas. Speaking to one of his Ju/'hoansi friends, he realises why his gift, given so grandly, had made him the target of mockery rather than gratitude:

> 'It is our way,' he said, smiling, 'We always like to fool people about that. Say there is a Bushman who has been hunting. He must not come home and announce like a braggard, "I have killed a big one in the bush!" He must first sit down in silence until I or someone else comes up to his fire and asks, "What did you see today?" He replies quietly, "Ah, I'm no good for hunting. I saw nothing at all [pause] just a little tiny

> one." Then I smile to myself,' he continued, 'because I know he has killed something big.
>
> 'Yes, when a young man kills much meat he comes to think of himself as a chief or a big man, and he thinks of the rest of us as his servants or inferiors. We can't accept this. We refuse one who boasts, for someday his pride will make him kill somebody. So, we always speak of his meat as worthless. This way we cool his heart and make him gentle.'

My colleague, Eleanor Power, has studied the fine line between gaining prestige and being accused of self-aggrandising behaviour in the rural south Indian villages where she works. Of particular interest is an annual celebration where villagers fulfil vows offered in thanks to Māriyammaṉ, a goddess who protects the village and its residents. Devotees are free to choose their vow – and the displays of devotion are accordingly variable. Many villagers join a procession, carrying clay pots laden with milk or fire pots full of hot coals, in a ritual designed to give thanks to the goddess. Others – typically men – fulfil more exacting vows, the most eye-watering of which is the *paṟavai kāvadi*. *Paṟavai* means 'bird', an apt descriptor for this particular act, where the devotee hangs from a crane by several meat hooks which are pierced through the dermis of his back and legs. Suspended thus, he is paraded through the village – his arms bearing fistfuls of leaves – while the crane swings him gently up and down.

The people who perform such acts of devotion stand to gain in *perumai*, which roughly translates as 'bigness' or prestige. But to gain *perumai* is to walk a metaphorical tightrope and, in Tamil, the word for being boastful or self-aggrandising – *tarperumai* – is closely etymologically linked to the first. When congratulated on their participation in one of these acts, devotees do not acknowledge their achievement, responding instead with the accepted refrain of 'it's just a vow'. Although there are reputation benefits to be gained from investing in these showy displays, Power's work also highlights the importance of investment into subtle social signals – regular worship at the temple, or small acts of kindness that may be seen only by the recipient – that allow individuals to forge and nurture important relationships without inviting accusations of boastful or grandiose behaviour.

Part 4

A DIFFERENT KIND OF APE

Until now, I've (mostly) treated humans as just another animal on the sprawling tree of life. We teach – but so do ants, meerkats and pied babblers. We punish cheats and care about our reputation – but so do cleaner fish. We live in tight-knit family groups, where individuals help one another to breed – but so do naked mole-rats and termites.

We may have an extraordinary amount in common with other species but, in this final part, I want to focus on the most important differences. As I've hinted in the previous chapters, it is not so much about what we do, but how we do it. Evolution finds many ways to reach the same behavioural destination and, in humans, it is often the cognitive routes we take to get there that set us apart.

In this final section, we'll discover how and why we became a different sort of ape. We are going to start by thinking about our unique evolutionary history, forged in the tempestuous climate of the Pleistocene, and how the intense requirement for cooperation also precipitated the great uprising of the masses against the classes. This, in turn, paved the way for the truly unique features of our psychology to emerge: the transition from despotism to a more egalitarian way of life meant that we started paying attention to what we had – and how that compared to the fortunes of others. We developed a sense of fairness.

This was a pivotal moment in the history of our species: forming coalitions helped us to reverse the predominant social order but created a different sort of pressure, where power and status became tied up in our social networks, rather than in our muscles. This means that, more than for any other great ape, our fortunes depend to a greater extent on the company we keep than on physical prowess. This social dependency has shaped our psychology, explaining many of our (otherwise quite strange) beliefs and even rendering us susceptible to mental disorders, such as psychosis.

14
FACEBOOK FOR CHIMPS

> I fully subscribe to the judgement of those writers who maintain that of all the differences between man and the lower animals, the moral sense or conscience is by far the most important.
>
> Charles Darwin, *The Descent of Man*, 1871

If someone were to ask you whether money can buy happiness, what would you say? The available data seems to support two contrasting positions. Population-wide surveys typically find that richer members of society are happier than poorer ones, suggesting that the answer might be yes. Nevertheless, although per-capita incomes in many Western countries have rocketed over the last fifty years or so, the average happiness of their citizens has barely changed, which seems to suggest that the answer is no. This apparent contradiction is what is known as the Easterlin Paradox, named after the professor who first described this strange pattern. The inconsistency can be explained as follows: it is not money per se that buys happiness, but knowing that we have more than people like us. By extension, the belief that you have less than your peers is one of the most significant causes of reduced life satisfaction in humans. Research done in Canada found that when people won the lottery, their neighbours were more likely to run up their debts and file for bankruptcy – ostensibly because they tried and failed to keep up with their lucky peers. Similarly, many Americans whose earnings place them in the top 1% (over $500,000 per year) say they are middle class, presumably because they compare themselves to those who are even richer.

And that's why Facebook for chimpanzees would never take off. It's not just that chimpanzees can't use computers or smartphones: Facebook is endlessly fascinating for many humans because it feeds our obsession with social comparison. Although we typically try to present the most flattering version of ourselves to others, it is now fairly well established that passively consuming the portrayals of other people's perfect lives has a damaging effect on well-being and mental health – particularly when our own lives appear to be less successful in comparison.

In fact, being worse off than peers is so unpleasant that we are willing to sacrifice our own resources just to stop a social partner from getting more. In the lab, the standard way to measure people's taste for fairness is to use the Ultimatum Game. This is a two-player game, where one person (the 'proposer') is given a sum of money and can offer a share of it to the partner (the 'responder'). The catch is that the responder has veto power: if they reject the offer, then neither player gets anything. If humans behaved in accordance with the models from classical economics, and simply tried to maximise their earnings in the game, then responders should accept any non-zero offer and, anticipating this, proposers should offer the smallest non-zero division of the pie. But, of course, this seldom happens.* Rather than being rational maximisers, most of us are irretrievably hung up on social comparison and fairness. Offers that are perceived

* Apart from among undergraduates studying economics. Various studies have shown that economics students behave more like 'economically rational' actors than do other undergraduates, both offering and accepting lower shares of the stake in the Ultimatum Game. It is not clear whether this propensity already exists among economists, or whether their education changes their behaviour.

to be unfair* are typically rejected, meaning that both players go home empty-handed. It may be economically irrational, but getting nothing is apparently more satisfying than allowing your partner to get more than you.

The taste for fairness seems to emerge quite early in human development. To measure it, researchers often use a currency that children care about and are highly motivated to get their hands on: sweets. In one experiment, conducted on children from around the world, pairs of children sat at opposite ends of a contraption bearing two trays of sweets, one for each child. One child – the 'decider' – had a pair of handles (one green and one red) that controlled which way the sweet trays would tilt. Pulling the green handle caused the sweet trays to tilt towards each child, pouring the sweets into each child's respective bowl. But pulling the red handle caused the trays to tilt away from the children, tipping the sweets into a central bowl that was then removed. Using this apparatus, the team could ask how children would respond when faced with unfair allocations. If the decider has one sweet on their tray, but the partner has four, would they accept, by pulling the green handle, or reject, by pulling the red handle?

Just like adults, children tend to reject inequity, forfeiting both their own and the partner's sweets just to stop the partner getting more. In the study above, older children were more likely to reject inequity, but even the youngest children in the study – the four-year-olds – were willing to pull the red handle if they were getting a bum deal. The fact that these patterns were observed across different countries, and in different kinds of society, suggests that disliking getting less than others – and disliking it enough to pay a cost to

* In Western societies, responders typically reject any offer that is less than 20% of the stake and, anticipating this, proposers usually offer around 40–50%. Nevertheless, there is considerable cross-cultural variation in what is perceived as being 'fair' and the corresponding offers made by proposers and thresholds at which people reject them.

stop it from happening – is something that is likely to be baked into the human psyche, rather than a preference that is learned during development.

But is this fretful tendency to compare our own fortunes to those of others a uniquely human foible? Or might it have deeper evolutionary roots? Although early work suggested that a concern for fairness might be something that we *do* share with other non-human primate species (and that social comparison is therefore something that might also have been present in the last common ancestor between humans and other apes), these results haven't held up terribly well when other researchers have tried to replicate them. In addition, many of the 'positive' results of inequity aversion in non-human primate species are also consistent with a simpler interpretation: subjects have expectations about the reward they will get, and are annoyed when they are presented with something inferior. A fairness preference is social by definition; it involves comparing your own pay-offs to someone else's. Non-human primates don't perform this social-comparison step, but instead evaluate what they get relative to what is theoretically available.

*

We seem to be the only primate species that frets endlessly over whether we are keeping up with the Joneses. Chimpanzees, on the other hand, couldn't give two pant-hoots about any of this. To understand why, we need to examine our own evolutionary history – and how the selection pressures that early humans faced might have differed from those of our great-ape cousins. It is hard to be certain, but the last common ancestor between humans and chimpanzees*

* When we talk about the last common ancestor of humans and chimpanzees, we are talking about the primate that was part of the family Hominidae, a creature who lived on Earth before members of the *Pan* and *Homo* genera, that gave rise to modern-day chimpanzees and humans, existed.

was probably more chimp-like than human-like, in terms of how they looked and how they lived. Learning more about chimpanzee society and behaviour is useful if we want to know more about what exactly it is that sets humans apart from our ape cousins.

We've already seen that the Western model of human society, with its emphasis on the nuclear family, is a spectacular outlier both cross-culturally and also in the broad historical context of the evolution of our species. For the vast majority of our time on Earth, humans would have lived in larger groups, where several families lived alongside one another. Like us, chimpanzees also live in multi-male, multi-female groups but, unlike humans, there are no family groups nested within them.

The classical view of ancestral (pre-agricultural) human societies is that they were small-scale, bounded communities, comprising just a few dozen members, with the idea being that 'each of our ancestors was, in effect, on a camping trip that lasted a lifetime'. But it turns out that this view is rather outdated. Humans were (much like we still are) likely to have been embedded in vast social networks, with many of their closest friends and family members living far away. Whereas the average male chimpanzee might expect to interact with just twenty other males in his entire lifetime, recent estimates put the average hunter-gatherer's social universe at about 1,000 individuals. Our earliest societies were therefore a unique mash-up between the nepotistic, extended-family units that we've seen in the other cooperatively breeding species, and the more fluid social worlds inhabited by chimpanzees, where individuals mostly mingle with non-relatives. But we also took things a step further than chimpanzees, extending the franchise beyond the immediate group, and drawing long threads between individuals living in one locality and friends and kin living elsewhere.

Life in a chimpanzee group is complicated: individuals have to manage many different social relationships, some of which are benign and others that are fraught with conflict. Males aggressively seek status, and the relationships they forge within their group can help them to achieve this aim. A crew of supportive henchmen helps an alpha male to secure his reign; while beta males can increase their chances of toppling the alpha male if they join forces with other ruthless characters. Alpha males cling to their throne with tight fists – with would-be usurpers watching closely for any signs that this grip is weakening. Those that succumb to illness, or sustain injuries, or simply grow too old, risk being aggressively and decisively deposed by coalitions of ambitious subordinates. Asserting dominance over other individuals in the group, and using coalitions as means to achieve this, are therefore likely to be traits that were shared by the last common ancestor with humans and chimpanzees – and are also present in our own species' psychological toolkit (and we will focus much more on our coalitional nature, and its implications for our mental health, in subsequent chapters).

The need to successfully navigate complex social worlds can drive the evolution of brain size and, like us, chimpanzees have large brains. Brains cost an energetic fortune, greedily creaming off around 20% of the body's energy stores, and costing around ten times as much, gram for gram, as skeletal muscle. The cost of this large brain is starkly highlighted by a traditional board game: chess. Two people sat stock-still, staring at a wooden board for hours on end might not seem to be on a physical par with a game of tennis or football. Yet, during a match, chess grandmasters experience elevated breathing and heart rates in the ranges normally seen in long-distance runners, meaning that they can burn up to 6,000 calories over the course of a single match. In 1984, the defending world champion, Anatoly Karpov, lost so much weight during the five-month final against Garry Kasparov that the match was unexpectedly – and controversially – halted by the competition's president.

Being brainy therefore means that you require a calorie-rich diet, and this is where we start to see points of divergence between humans and other apes. Our great-ape cousins that are alive today live in relatively forested and aseasonal environments, where food is readily and predictably available. For instance, gorillas essentially live in giant salad bowls, and can subsist on leaves and other woody materials from the trees around them. A chimpanzee diet is richer, comprising mostly ripe fruits – but these are still relatively easy for a single individual to find for himself. Humans are different. For one thing, we have much larger brains (relative to our body size) than chimpanzees or gorillas. As we saw earlier, we also evolved in environments where feeding these onboard computers would have been more difficult. For most of our time on Earth, humans have eked out an existence in an environment that was both dry and unpredictable and where most of the food we ate had to be hunted, searched for or scavenged. To find food, we had to work together and we needed one another in a way that chimpanzees don't.

Given the extent to which each individual's own survival and reproductive success was so dependent upon the efforts of others, it starts to make sense that humans would have evolved a suite of socio-cognitive traits dedicated to monitoring and evaluating interaction partners, as well as presenting ourselves in the best light possible: a reputation for fairness would have helped us to attract partners to undertake risky collaborative ventures that were essential for survival. Interdependence also helps to explain how humans became so much more effective at hunting than chimpanzees. If all it took was raw muscle power and brute force, then chimps should be able to wipe the floor with us. But chimpanzees seem to be lacking in a crucial area, one where humans excel: collaboration.

Chimpanzees are known to hunt and eat red colobus monkeys. Although a solo male typically initiates a hunt, others often join in; and hunting success is much higher when chimps hunt as a group rather than individually. During the hunt, chimpanzees adopt different roles: one male might flush the monkeys from their refuge, while another blocks the escape route. Somewhere else, an ambusher lurks, ready to make his deadly move. Although this sounds a lot like teamwork, recent work offers a simpler interpretation. Chimps are more likely to join others for hunts because larger hunting groups increase each *individual's* chance of scoring a monkey – they aren't interested in collective goals. The appearance of specialised roles in the hunt may also be an illusion: a simpler explanation is that each chimp places himself where his own chance of catching a monkey is highest, relative to the positions the others have already assumed. Whereas collaboration in chimps seems to emerge from an 'every chimp for himself' mentality, humans seem to do things differently, being wired for teamwork in a way that chimps are not. For chimps it's 'me' but for humans it's 'we'.

If given the choice to work together to achieve a reward or to work independently, human children tend to prefer the collaborative option. Chimps, on the other hand, prefer to work alone. The reason for this difference seems to be related to the hierarchical nature of chimp societies, and the fact that chimpanzees don't like to share. A subordinate who works with a dominant to secure a joint prize may well end up with nothing to show for it, because the higher-ranking chimp can, and probably will, steal the prize for themselves. A chimpanzee who catches a colobus monkey will typically try to monopolise this food package for himself. Although small scraps may be relinquished to important allies, or to potential mating partners, most of these transactions involve meat being begrudgingly taken from the prize-holder or handed over in response to persistent badgering from whoever is closest. There are very few reported cases of chimps voluntarily sharing food, in the absence of begging or harassment from others.

Teamwork in humans, on the other hand, is scaffolded by meritocratic principles. In foraging societies, hunters share meat according to who helped to catch it and the importance of their role. Among Efe pygmies, the hunter who shoots the first arrow gets most of the meat, followed by the person whose dog chased the prey, and so on. Among Dominican fishermen, the proceeds from fish sales are used first to compensate the boat owner for the petrol, with the remainder subsequently being divided equally among all crew members. In experiments where toddlers work together to obtain rewards, they also willingly share these after the fact. By contrast, chimpanzees don't seem to care about compensating individuals who play a role in joint success. Although recent work on chimps from the Taï National Park in the Ivory Coast hinted at this possibility, follow-up experimental work on captive chimps suggests that being closest to the meat-holder is the main predictor of who might get a share of it. Being in the right place at the right time, rather than participating in the collaborative venture, seems to predict whether chimps get a share of a food prize. Given how little there is to be gained from collaboration, perhaps it is not so surprising that chimpanzees often prefer solo pursuits.

*

Collaboration relies on communication – and this is another arena in which we stand apart from the other apes. It's not just that we have language (although this undoubtedly helps): even if we restrict ourselves to forms of non-verbal communication, such as hand gestures and eye gaze, we see that humans use and understand these in vastly different ways to chimpanzees. The key difference seems to be that humans understand gestures as signals that are designed to help them, whereas chimpanzees do not. In an experiment where children have to search for a hidden toy, they intuitively understand that an adult pointing to – or looking meaningfully at – an overturned bucket is informing them of the reward's location. One can immediately envision how the ability to send and receive these signals could be helpful in the context of collective action, such as hunting. Pointing and looking could be used to signal the whereabouts of the prey, or where the other person should position themselves without revealing the location of the hunters to the intended prey. Intriguingly, domestic dogs outperform chimpanzees on tasks like this – being able to understand a wide range of human signals as being informative and cooperative. Wolves do not share these abilities, which suggests that dogs have co-evolved these socio-cognitive traits because of their long-running partnership with humans.

Chimpanzees face a fundamental barrier to using these helpful signals because they don't instinctively understand them as something that is designed to help or inform. In the same experimental paradigm as above, where the experimenter points to the hidden location of a reward, chimpanzees struggle to infer that the experimenter is trying to show them the location. For chimps, 'you're trying to help me find something' is a conclusion that is seldom reached. It is not because chimpanzees don't use gestures; indeed, they have a far richer repertoire of gestures than even we do (perhaps because many human gestures have been replaced by language). But these are used to get other individuals to do what they want (e.g. 'give me some of that', 'bugger off', and so on) rather than to helpfully direct another chimp's attention to a goal or to indicate a shared interest in something.

Pointing is a case in point. Although they don't typically do this in the wild, human-raised chimps can learn to point. Unlike human infants, however, chimpanzees never point to highlight something interesting, but only to indicate that they want something. Proto-declarative pointing – the kind that children do when they want you to look at a plane or a bird or anything else of interest – is a key developmental milestone and something that most infants spontaneously do by the time they are about nine months old. Infants who do not point in this manner – to share an interesting observation about the world with an adult – are also typically more at risk of later being diagnosed with autistic spectrum disorders, where a feature is the reduced tendency to consider the mental states of other individuals. Pointing at something to show it to someone relies on at least a basic appreciation that other individuals have mental states, and that these are something that you can influence. The 'hey, give me that' (proto-imperative) pointing, on the other hand, doesn't require you to understand or share the mental states of other individuals, but instead could be reinforced through simple trial and error (e.g. if I point at the thing I want, I'm more likely to get it).

One general conclusion from all these disparate lines of research is that humans think in terms of 'we' whereas for other apes, it is more likely to be 'me'. Humans don't like getting less than other people in social interactions, especially when we played a part in securing the resource at stake, whereas chimps don't engage in social comparison. Our sensitivity to fairness, and to the intentions of our interaction partners, likely shaped some of the unique aspects of human cooperation, including the tendency to genuinely care about the welfare of others and to voluntarily help them, at a personal cost to ourselves.

In fact, the human desire to help others seems to be present in our cognitive toolkit from as early as we can measure it. For instance, from around eighteen months old, human infants will spontaneously help an adult to achieve a goal, even without being asked and when there is no reward for doing so. If captive chimps are exposed to the same paradigm as these toddlers, they do occasionally help but nowhere near as frequently as the children. Other work indicates that chimps really care very little about the welfare of social partners. In an experiment where chimps could decide whether to choose an option that delivered a food reward to them *and* a familiar partner, or an option that delivered a reward to just them, chimps picked either option at chance. Like other non-human primates, chimpanzees don't seem to care much about what other individuals in their group get – they simply focus on themselves.

*

Our ability to collaborate made us deadly hunters. But the development of our hunting skills had another, unforeseen consequence: it allowed us to subvert the social order. A dominant male chimp rules by force, relying on physical prowess to throw his weight around. Dominance is taken and imposed upon others, against their will. But ancestral humans did things differently. In a world where anyone could hurl a stone or strike a spear from afar, brawn was no longer a viable means to wield authority, because even puny enemies could fight back. The mutiny had begun.

15
MUTINY

Man is by nature a political animal.

Aristotle, *Politics*, 350 BCE

On a wintry Sunday in 1787, HMS *Bounty* left Portsmouth on an audacious mission: to circumnavigate the southern half of the globe. The crew were headed for Tahiti, where they intended to gather breadfruit trees and to transport them across the seas to the West Indies. It was thought that the plants might grow well in this British colony – and that the fruits could be used a cheap means to feed the people the British had enslaved on these islands.

The *Bounty* made it to Tahiti but it never completed its onward journey. After five months living on a tropical island, the crew were disinclined to return to the rigour and discipline of life at sea. Although the journey to Tahiti had been largely free from strife, William Bligh, the ship's captain, now resorted to increasingly punitive tactics to force his recalcitrant crew to toe the line. Floggings, humiliation and the withholding of rations became commonplace, and the mood on the ship soured. Barely three weeks after they had left Tahiti, the crew rebelled. In the dead of night, a handful of men led by Fletcher Christian, Bligh's former right-hand man, captured the captain and threw him onto a rowing boat with a handful of loyal men. Tossing in some cutlasses and just enough food and water to survive for a few days, the mutineers cast their former colleagues and friends adrift.

The *Bounty* may be the most famous tale of mutiny, but the threat of rebellion was ever-present aboard merchant ships of the eighteenth and nineteenth centuries, and likely stemmed from the way that these operations were run. The vessels were typically owned by rich traders, who did not wish to undertake lengthy and perilous journeys across the oceans themselves. But entrusting their ship and precious cargo to a leaderless mob of seamen for months at a time was also unappealing. How could they ensure that the crew would take good care of their assets? What was to stop the men from appropriating the vessel and any goods, or dishonestly reporting the sums of money they earned and keeping larger profits for themselves?

One solution to this problem was to appoint a single leader of each ship: a captain, like Bligh, whose interests were financially aligned with those of the ship's owner, and who was given the authority to rule with an iron fist. If a crew member misbehaved, the captain had licence to administer various forms of punishment, an entitlement that many captains abused. The economist Peter Leeson (in what must be the best-titled academic paper of all time[*]) describes how the ships' captains – corrupted by power – frequently became 'predatory'. Predatory captains made their subordinates' lives miserable in countless ways, including by forcing the crew to undertake voyages to places that they had not been contracted to visit, restricting rations, docking or withholding wages, tossing them overboard, and beating them (sometimes to death). Small wonder, then, that this period became known as the great age of mutiny.

But mutiny wasn't the only concern: this was also the golden age of piracy. Encountering a pirate ship on the high seas must have been terrifying: pirate crews were gargantuan, frequently numbering in excess of 150 men[†] compared to the twenty or so sailors aboard the

* 'An-*arrgh*-chy: The Law and Economics of Pirate Organizations'.

† Although most pirates are men, there are a few notable exceptions. Indeed, the pirate broadly considered as having been the most successful in history (Ching Shih) was a woman.

merchant vessels. One might presume that controlling such a large band of vagabonds and rogues would be far more difficult, and that mutiny would be rife aboard pirate ships. But the reality was quite the opposite: for the most part, pirate life was actually more peaceful and less anarchic than on the merchant ships.

Pirates seem to have solved the problem of mutiny by placing authority in the hands of the many, rather than the few: in other words, by inventing their own form of democracy. It helped that pirate ships were typically co-owned by all crew members, so all men were incentivised to take care of the ship. Although pirate ships also had captains, their remit was far more circumscribed. Pirate captains acted as decision-makers in times of battle but in all other matters were on an equal footing with the rest of the crew. Crucially, they were democratically elected and a captain was apt to be demoted if he was not seen as serving the best interests of the crew.

Another key ingredient in pirate democracy was the separation of control of the crew from control of the treasure. The shares of any profits gleaned from plundering another ship were determined by another democratically elected pirate, called the quartermaster. As with the captains, any self-serving quartermaster could also be demoted. To keep the peace on pirate ships, all members were forced to comply with a written and agreed-upon set of rules: a constitution of sorts. Pirate constitutions included rules that helped keep life on the ship tolerable and peaceful, including directives such as 'The lights and candles to be put out at eight o-clock at night' and 'No striking one another on board, but every man's quarrels to be ended on shore, at sword and pistol.' Importantly, these rules also described how treasure gleaned during a raid ought to be distributed: individuals who took greater risks during battle, or who incurred a severe injury, were entitled to a proportionally larger share of the booty. These three cornerstones – democracy, separation of power, and formal, legitimate constitutions – helped to keep the peace aboard pirate ships in a way that tyranny and despotism on merchant ships simply couldn't.

*

The contrast between the kind of social organisation that emerged aboard the merchant ships and the kind that pirates forged for themselves illuminates some rather general features of human nature. The desire to predominate over others, to try to appropriate slightly more than our fair share, and to seize power where we can, are all fundamental aspects of human nature. Nevertheless, cultural inventions, like the democratic systems invented by the pirates, can hold the most ruthless individuals in check. These tales of revolt and rebellion also reveal something more primal about our social lives: in watching these conflicts play out, we see the extent to which power and status depend upon the support of others in our social network. Friends really do come with benefits.

The unromantic truth is that coalitions, friendships and alliances function as social tools that help us to achieve our goals (even if we don't consciously think of them in this way). In Chapter 10, we saw how friends might provide mutual aid to one another, which is especially important when resources come and go in an unpredictable manner. Remember the *osotua* friendships in the Maasai, where herders offer 'no strings attached' help to their most important partners on the understanding that they will receive similar assistance should they ever require it. Other work has shown that the support conferred by friendships can improve happiness and wellbeing, lower stress, boost immune function and even help people to live longer. Experiments done with the BaYaka hunter-gatherers of Congo and Central African Republic indicate that people with more friends have higher BMI, which translates into increased fertility for women. In this society, men with the strongest social networks have an increased chance of taking additional wives, directly impacting reproductive success. Friends are especially important when people face adversity: in Andersonville, a Confederate prisoner-of-war camp in the United States, nearly half of all prisoners died. In this terrible environment, one of the strongest predictors of survival was having a friend.

As well as helping us to weather adversity, friendships can also help us to cement or improve our social position, by offering protection or by improving our ability to compete with others. Female baboons benefit when they form friendships with unrelated males because these allies help to protect the female (and her offspring) from other infanticidal males within the troop. In chimpanzees, friendships seem to be especially important for males, where they can help individuals to improve their rank or to gain access to females. An alpha male's tenure frequently relies on the support of a close ally, who helps the alpha fend off attacks from competitors, in exchange for tolerance and a share of reproduction.

But not all friendships last. Coalitions thrive on conflict and, just as conflicts of interest can change among different individuals or groups, so too can the strength of social ties. Remember that the man leading the mutiny against Bligh – Fletcher Christian – was formerly one of the captain's closest allies aboard the ship. On a previous voyage, Bligh had granted Christian privileges normally reserved for higher-ranked seamen and, on the *Bounty*'s outward voyage, had even controversially promoted Christian above a more experienced crew member. And yet it was Christian, his protégé and friend, who organised the insurgence and cast Bligh adrift in the Southern Ocean.

A series of observations made upon a population of chimpanzees in Tanzania in the early 1970s illustrates how shifting friendships can affect fortunes. The conflict at the heart of this story centres around three male chimps. The alpha male, Kasonta, was a 'very aggressive and powerful' chimp – traits that enabled him to rule the roost for the best part of six years. Throughout this time, he faced multiple challenges from Sobongo, his immediate subordinate. Kasonta was only able to maintain his alpha position for such a long time in the face of these constant challenges because he had the support of a third, lower-ranking male, Kamemanfu.

Unlike the others, Kamemanfu was a smaller chimp – he could not hope to secure access to females by fighting his way to the top. But what he lacked in brawn he made up for in brains. As time went on, Kamemanfu fuelled the conflict between the warring males to his own advantage. Rather than being beholden to either male, Kamemanfu's strategy was one of fickle allegiance, where he switched his support from one male to the other. This destabilised the hierarchy and exacerbated the conflict between the two rivals. Given the enmity between them, neither male could afford to alienate Kamemanfu: a potential ally could make the difference between winning and losing the fight. Through this divisive strategy, Kamemanfu bought the tolerance of the more powerful males, earning himself the right to mate with females in the group in exchange for his support, however capricious it might have been. This study is admittedly anecdotal in nature, but subsequent studies have confirmed the importance of such political manoeuvrings in dictating the rise and fall of leaders within chimpanzee societies.

*

In a world where allegiances can change, and former friends can become foes, individuals need to remain hyper-vigilant to cues of threat from other people ('social' threat), monitoring the relationships between others in the group – and even disrupting these if they might pose a problem in future. For instance, a recent study of wild chimpanzees found that bystanders frequently intervene in the grooming interactions of other individuals, not because they wish to be groomed themselves, but to disrupt the nascent bonds that form during these intimate encounters. Bystanders were especially likely to intervene when one of the grooming individuals was a close affiliate (suggesting that bystanders wish to monopolise the support they receive from these friends) and when both the grooming individuals were lower rank because, as the previous example showed, an allegiance of two low-ranking males can have devastating consequences for a more dominant individual.

Humans are especially concerned with these details of social life. We reflexively categorise others as 'in-group' or 'out-group', even on the basis of completely arbitrary cues (early experiments found this groupish effect could be induced via the colour of name badges, or the preference for paintings by Picasso versus Monet). Our excessively groupish psychology is – ironically – a consequence of our exceedingly cooperative natures. By working together, the earliest humans were increasingly able to overcome the challenges that nature threw at them: the problems of food scarcity, water shortages and dangerous predators could all be mitigated via cooperation. But, as a consequence, other humans became our primary threat. We were no longer battling against nature, but against each other.

In this context, evolution would have placed a high premium on social competence: the ability to nurture and curate our own social support network, to monitor the friendships and alliances of others we encountered, and – most crucially – to detect and avoid social threat. When these threat-detection systems work well, they keep us safe from danger. But when they fail, we risk becoming a danger to ourselves.

16
HERE BE DRAGONS

Even paranoids have enemies.

Golda Meir, 1973

In one of the earliest documented case studies of what would now be diagnosed as schizophrenia, Bedlam patient James Tilly Matthews described being persecuted by the 'air-loom gang' of which he could name the individual, but illusory, members. According to Matthews, this gang of criminals and spies had access to an 'air-loom', a ray-emitting machine that gave them various special powers, including the ability to perform 'lobster cracking' (interrupting the blood flow inside his body, leading to an instant painful death) and 'thought making' – the ability to insert and control Matthews' thoughts, as well as those of leading government figures at the time.

The troubling thoughts experienced by James Tilly Matthews are a common and often debilitating feature of psychotic-spectrum disorders* and, in fact, the themes of Matthews' delusions – the idea that others could control his thoughts or harm him from afar – are common among patients suffering from such disorders. Although Matthews' beliefs were unusual in terms of their rather bizarre

* Psychosis is not to be confused with psychopathy. Psychosis occurs when you lose your grip on reality. Two of the most common symptoms of psychotic-spectrum disorders are hallucinations (usually in the form of hearing voices) and paranoid thoughts or delusions.

content, the underlying paranoia upon which these delusions are built is actually rather common in the general population, where it occurs across the full spectrum of severity and need not be indicative of any mental disorder. In everyday language, it is common for people to speak of being paranoid in scenarios where they were actually worried or anxious ('I was really paranoid I was going to catch that vomiting bug!') but in a clinical sense paranoia has a different meaning. Although worry and anxiety involve the belief that something bad might happen, paranoia has an additional feature: believing that harm is *intended* by another person.

Paranoia most often presents in the form of mildly exaggerated socio-evaluative concerns (such as believing that others are gossiping about you), or as feelings of mistrust or suspicion. However, in more extreme cases, as with James Tilly Matthews, it can become pathological. We can measure the proneness for paranoid thinking across the full spectrum using questionnaires where people are asked to what extent they agree with statements about the harmful intentions of others. These statements include things like 'I was convinced that people were singling me out', and 'Certain individuals have had it in for me'. In a sufficiently large general population sample – 1,000 people selected at random – more than half of participants would experience paranoid thoughts either infrequently or not at all. Around 150 (15%) of participants might regularly experience feelings of mistrust or suspicion, and around thirty or forty of them (3–4%) would be quite preoccupied by these paranoid thoughts, scoring in the same range as people who have been diagnosed with a psychotic-spectrum disorder.

One way to view paranoia is solely as a pathology: an unwanted symptom of a mental disorder that should be treated and, ideally, eradicated. Over the last few years, I've worked with Vaughan Bell (a clinical psychologist at UCL) to offer a different perspective: paranoia might be a feature, rather than a bug, in our psychology. We are emphatically *not* proposing that the extreme paranoia that accompanies mental disorders like schizophrenia has been favoured by evolution. Although paranoia is part of a normally functioning human psychology, it is undeniable that for a small subset of people, the frequency and severity of paranoid thoughts tips over into what is clearly a troubling and maladaptive outcome for the sufferer. At lower intensities, however, paranoia is likely to play an important role in helping us to detect and manage social threat.

To understand how paranoia can be mostly protective and yet occasionally harmful, analogies with other protective mechanisms might be helpful. Fever is a good example. The ability to raise core body temperature is a vital part of an immune response, which allows the body to fight off infection. But occasionally fever itself can become dysregulated, with potentially life-threatening consequences. Similarly, pain is an unpleasant and distressing thing to experience but it alerts us to physical harm, prompting us to act to avoid further damage. Occasionally, pain can also continue to be experienced long after the initial injury has resolved, resulting in chronic pain disorders where the pain has no benefit. Paranoia can be understood in similar terms: for most people, most of the time, paranoia is a mechanism that works as it should and serves a protective function. But for a minority, things clearly go badly wrong.

So, what is the potential advantage of being mistrustful or suspicious of others? As we know by now, humans evolved in complex groups where some of the most significant threats were posed by other humans. In such environments, selection should favour cognitive mechanisms that allow individuals to monitor the potential for threat and respond to it, by avoiding or incapacitating malevolent others. Developing a brain that has a capacity for paranoid thinking is one way that evolution helped us to achieve this. Importantly, a proneness towards paranoia could be favoured by selection even if it regularly produces false alarms. This is what's known as the smoke-alarm principle: it's better to be safe than sorry. The smoke-alarm principle implies that a capacity, and maybe even a tendency, towards paranoid thinking is a vital part of a well-functioning cognitive toolkit – something that helps to keep us safe from people who might wish to harm us.

An evolutionary perspective also suggests how paranoia might manifest. Specifically, paranoia should be flexible and attuned to features of the social environment, working a bit like a volume dial. Paranoia should increase when perceptions of social threat are high and decrease when perceptions of social threat are lowered. You can think of this as being a bit like the relationship between a glucose deficit in your body, and the feeling of hunger that typically ensues. Our bodies continuously monitor internal glucose levels, and we experience variation in the changing internal environment through the subjective feelings of hunger and satiation. Humans also have a 'social-safety index' that functions in a similar manner: it tracks the social environment and monitors cues of social support and social threat. The subjective safety estimate that is produced can manifest as paranoid thoughts and feelings that we all experience from time to time.

Just as hunger can be exacerbated or alleviated by fasting or eating, the social-safety index is expected to respond to changes in safety or threat from other people. Those who are consistently exposed to cues of social threat are expected to experience more paranoid thoughts in their daily lives. Large-scale epidemiological studies examining the risk factors for developing psychosis (of which paranoia is a characteristic feature) support this idea: although psychosis has a genetic component,* environmental factors also play a large part in determining a person's susceptibility. A closer look at the kinds of environmental factors that predispose people to psychosis or paranoia underlines the importance of exposure to social threat. People are more likely to report experiencing paranoid thoughts if they have a history of being bullied or

* As with nearly all phenotypes, psychosis risk is associated with many genes of small effect.

victimised, or if they have a small social network (which is a proxy for low social support). Similarly, being low status is a risk factor for developing psychosis, as is being part of a marginalised ethnic or religious group. In one area of south-east London, the incidence of psychotic-spectrum disorders was found to be three times higher among Black individuals than among the white people living in the same neighbourhood.

Feeling marginalised, or being low status or having a small social network, are things that should set the social-safety alarm bells ringing. The otherwise paradoxical finding that the psychosis risk experienced by ethnic minorities can be buffered when these same minorities live at higher densities with their ethnic group (the so-called 'ethnic-density effect') is also exactly what we might expect based on a social-safety account of paranoid thinking, because there is increased safety in numbers than in isolation. In the south-east London study, the link between psychosis and race was no longer apparent in the communities where Black people comprised at least a quarter of the population.

But, as suggestive as these epidemiological studies are, they cannot tell us anything about the causal relationship between social threat and paranoid thinking. In other words, these kinds of studies can't tell us whether experiencing social threat *causes* people to become more paranoid, or whether things might be the other way around. One way that the causal relationship could conceivably be reversed is if more paranoid individuals tend to withdraw from social interactions, which causes their social networks to shrink. If so, a small network size would be a consequence rather than a cause of variation in paranoia.

To establish causal relationships, we need experimental approaches where social threat can be manipulated, and the propensity to engage in live paranoid thinking can be measured. To that end, we have developed experimental approaches that allow us to not only measure, but also *manipulate,* paranoid thinking among members of the general population. If paranoia is the output of a cognitive mechanism that monitors social threats, then it should also be flexible, meaning that we should be able to turn the dial up or down in a laboratory setting.

We've measured this by exposing people to mild social stressors in genuine social interactions with others. The experimental paradigm we use is the Dictator Game: a two-player interaction where one person (the 'dictator') is given some money and can choose whether to send half or none of this money to the partner (the 'receiver'). In this paradigm, the receiver is totally powerless and must accept whatever the dictator sends to them (hence the name of the game). In most Dictator Game experiments, the focus is on the dictator.* By contrast, in our experiments, we care much more about the receiver's perspective. We want to know how people attribute intentions to a dictator, when the reasons this person sends (or refuses to send) money are quite difficult to discern.

To elicit intention attributions, we ask people how they interpret the actions of others. After finding out how much the dictator sent to them, the receiver is asked to infer the motives underlying the dictator's decision to send (or withhold) any money. Importantly, the dictator's true motives are highly ambiguous: someone who keeps the money for themselves might just be greedy, which would imply that they are motivated purely by self-interest. But, a receiver might also believe that the dictator kept all the money specifically to deprive them of getting anything: receivers might therefore believe that their partner was motivated by harmful intent. The ambiguity in this setting is crucial because it maps onto our real-world experience, where people's true intentions are unknowable most, if not all, of the time. Where there is ambiguity, there is also scope for

---

* This is because Dictator Game-giving seems to undermine the rational-actor model of human behaviour, that predicts individuals will keep all the money for themselves.

variation and error in interpreting the actions of others. In other words, this zone of uncertainty is the reason that people can interpret the same actions in vastly different ways.

In our studies, most people infer that a selfish dictator (who keeps all the money for themselves) is simply greedy. Relatively few people believe that the dictator is also malevolent. As you might expect, however, people who are more prone to paranoid thinking in their day-to-day lives are more likely to attribute harmful intent to the dictator, with the severity of their paranoia predicting the degree of malevolence they attribute. This much is unsurprising. The real test of our thinking hinges on our ability to *manipulate* people's tendency to make paranoid attributions about their partner by baking social stressors into the experimental setting.

We've done this in a few ways. For example, we can ask participants to tell us something about themselves, such as whether they are politically conservative or liberal; or whether they perceive themselves to be relatively high status or low status, compared to other people in society. We can then use this information to manipulate the social threat experienced in the Dictator Game. Our studies were run using participants from the US – and so we asked them to indicate whether they identified more strongly as a Republican or a Democrat. Knowing that your partner is a political opponent is mildly socially threatening, compared to being paired with a politically like-minded individual. Similarly, being told that you are interacting with someone who is higher up the social ladder should be more threatening than being paired with someone who is on the same rung, or even below you.

As expected, we routinely find that exposing people to these social stressors turns the dial on paranoid thinking: people are more likely to believe that the partner has harmful intentions. The implication is that, rather than being an immutable and broken feature of people's brains, paranoia is better understood as being a flexible – and perhaps beneficial – response to social threat.

*

One of the commonest ways that paranoia manifests in the real world is in a tendency towards believing conspiracy theories: the sense that groups of powerful agents are acting with nefarious goals in mind. The mistaken belief that vaccines cause autism, for example, is a primary cause of the current upsurge in cases of measles around the world, with the World Health Organization recently reporting that there were more cases in the first half of 2019 than in any other year for over a decade. Similar conspiracy theories about vaccines against COVID-19 are already emerging, and will likely represent a significant obstacle to achieving herd immunity to this disease via vaccination. Often, conspiracy theories are just plain weird. For example, the 'Lizard People' theory describes the conviction that various famous figures of authority (including George W. Bush and Queen Elizabeth) are in fact lizards, that assume an impeccable human disguise in their pursuit of world domination. (Apparently, the telltale signs of a Lizard Person are green or blue eyes, abnormal-sized pupils and low blood pressure, which are all traits that I possess.)

Conspiracy thinking is not a niche sport: more than half of all Americans endorse at least one conspiracy theory – with the tendency to believe in one kind of conspiracy theory meaning that you are also more likely to believe another. Despite the folk wisdom, conspiracy thinking is not associated with being on the left or the right of the political spectrum; instead, it is people at the extremes of these ideologies who are the most likely to harbour these beliefs. Furthermore, conspiracy thinking is not a New Age malady that we can attribute to the rise of social media or other novel ways that information and rumour can spread. A study measuring 120,000 letters written to the editors of the *New York Times* and the *Chicago Tribune* from 1890 to 2010 found the prevalence of letters espousing conspiracy theories to be largely consistent over time. The tendency to suspect that bad actors are finding ways to subvert social order for their own malicious ends isn't a recent invention, but is a part of the collective human mind.

But this seems to present a paradox: how can we simultaneously be a species that can split the atom and land spaceships on the moon, and yet be so collectively vulnerable to what many people think are irrational, unfounded and nonsensical beliefs?

Addressing this requires us to take a step back from the issue and ask what our beliefs are really for. We tend to think of our own beliefs as akin to maps: representations that provide accurate information about the world. According to this view, beliefs exist to help us effectively navigate through life, enabling us to choose the appropriate behaviours to reach the desired destination. Although some beliefs obviously do work in this way, much – perhaps even *most* – of what we believe has little or no basis in reality. We readily adopt beliefs that exist only in our shared imaginations and that are not easily provable or based in anything we can point to as a verifiable fact in the world. We believe that there is a border between England and Scotland, but if all evidence of human activity was wiped from Earth, could we show an alien observer where this border was? Most people think that money carries extrinsic value, but could we convince the alien that a £10 note – a scrap of paper – was something of real worth? A sizeable fraction of the world's population worships an omnipresent supernatural being, who watches our every move and stands ready to throw us into the hellfire of eternal damnation if we err. What, if anything, is the difference between a belief in God, and the belief that you are being persecuted by an air-loom gang, or that the CIA have turned your bones to steam?

*

Part of the answer to this might lie in the fact that most of the 'strange' things we believe are beliefs that we share with others, an insight that can help us to distinguish conspiracy theories from frank persecutory delusions. Unlike delusions, which are typically privately held beliefs that the individual in question is being personally targeted, conspiracy theories are usually entertained by a larger group of individuals, and involve the perception that powerful agents are acting with nefarious goals in mind, but not that the believer themselves is being personally targeted.

Truly delusional beliefs strike us as odd precisely because no one else endorses them. This suggests that we sort beliefs into two piles – justified and unjustified – not on the basis of a rational, scientific method but through a sort of social valuation process. An implausible belief is deemed 'rational' when sufficient others believe it and 'irrational' or even crazy when it is held in isolation.

In one sense, therefore, beliefs function like institutions: the societal rules and norms of interactions, which are in an individual's best interest to follow if they are adopted by the majority. It is rational to believe that money has a value, if that is what everyone else in your society currently believes. It is when the collective confidence in this belief wavers that the value of currencies can collapse. This suggests a role for beliefs as helpful fictions, allowing us to form collective agreements about the way things work – and to engineer better solutions to the game of life we all play.

In other cases, however, beliefs can become tags for group membership. The belief that anthropogenic climate change is a hoax is associated with a more conservative world view, while the belief that there are no biological differences between the sexes represents a more left-wing stance. Although there is no clear consensus as to why particular beliefs become attached to specific groups, once it does happen, the endorsement of that belief can act as a credible signal of an individual's commitment to their group. This means that it can be beneficial to hold a belief – even a false one – if that is a prerequisite for membership of a group, or predicts the level of support you will receive. Holding the socially sanctioned belief unlocks a flow of social benefits, whereas defending 'prohibited' beliefs can result in costs: derogation, exclusion, ostracism, even murder. Over history, countless people have been imprisoned or executed for religious beliefs or practice, for holding beliefs that contradict some other version of reality. More recently, we can observe similar trends on social media platforms like Twitter, where people risk being 'cancelled' for airing views and opinions that clash with the majority view.

The idea that beliefs function more as signals of group membership than as vessels of epistemic truth might help us to understand why our brains seem to be chock-full of software that enables us to defend these ideas, even in the face of countervailing evidence. This can happen consciously or subconsciously; and can infiltrate cognitive processes as diverse as perception, memory and evaluation of evidence. For instance, in one study, a person's political identity affected their perception of the colour of Barack Obama's skin, with Republicans perceiving Obama as having darker skin than did Democrats. Trump (versus Clinton) supporters were more likely to mistakenly identify a picture of Barack Obama's inauguration crowd as the 2017 inauguration of Donald Trump, while Democrats were more likely to misremember George W. Bush as being on holiday during the Hurricane Katrina crisis. Conservatives and liberals, who exhibited no differences in their ability to solve a maths problem when neutrally framed, struggled when the same puzzle required them to conclude that gun control either prevents crime or has no effect, respectively.

The cognitive quirks that seem to give rise to 'irrational' beliefs or faulty logic – confirmation bias, motivated reasoning, selective memory, and so on – can therefore potentially be understood as adaptive proximate mechanisms that help us to adopt and defend beliefs that will be most beneficial in our current social setting, rather than the beliefs that are objectively correct. This proneness to believe what we cannot prove and for narratives to shape how we see the world is a crucial part of the human success story. But there is the potential for much harm. On an individual level, when these processes go awry, they can manifest as the experience of intense paranoia and delusional thought. On a societal level, the threats may be existential. The social nature of belief prevents us from taking someone else's perspective, generates confusion over the nature of true and 'alternative' facts, and can sow discord and conflict in situations where solidarity is needed. To give a recent salient example, at the start of the coronavirus pandemic, some conservative-leaning media outlets peddled misinformation that COVID-19 was no deadlier than seasonal flu, a conspiracy narrative which gained greater traction among political conservatives than among liberals. By November, the consequences of these misguided beliefs were apparent. A US-based study, which used location data from 15 million smartphones per day, found that consumption of conservative media was linked to measurable differences in physical distancing – with inevitable downstream consequences on infection and death rates in conservative-leaning regions.

*

In this chapter, we've seen how coalitions shaped our psychology, potentially introducing the human susceptibility to psychotic-spectrum disorders. We've also seen that coalitions shape our perceptions and beliefs. In the next chapter, we'll see that coalitions did more than this: they changed the very fabric of the societies we live in. Humans are great apes, meaning that we come from a long lineage of species where social order is built on despotism and hierarchy. But our coalitional natures allowed us to do something quite extraordinary: by joining forces, we managed to turn the tables, taking the power from the hands of the few and distributing it among the masses.

This subversion of the social order is what the anthropologist Christopher Boehm has called the 'reverse dominance hierarchy'. Instantiating it is arguably the single biggest factor that set our own species on its own unique path. Reversing the social pecking order relied on our ability to form large 'macrocoalitions': group-level alliances that prioritise the interests of the collective. By acting in concert to thwart the ambitions of tyrants and bullies, macrocoalitions levelled the playing field and prevented powerful elements within the group from rising too insistently to the top.

17
TAKE BACK CONTROL

> United we stand, divided we fall.
>
> Aesop, 'The Four Oxen and the Lion'

From our modern vantage point, any attempt to characterise human society might appear to be a fool's errand. Our social worlds are a constellation of possibilities, ranging in size from nomadic bands of a few dozen individuals to nation states of billions; some lack any formal leaders, while others are ruled by chiefs or dictators or democratically elected governments. Take a moment to consider just how unusual this makes us. Elsewhere in nature, universality of social arrangements is the norm. For instance, although the babbler groups I worked with all lived in South Africa's Northern Cape, if I had travelled to Botswana, or Namibia, or even Zimbabwe, any pied babblers I found there would have lived similar lives, in groups of more or less the same size, and with the same kind of hierarchical relationships.

Although our societal diversity is unusual, this is probably a state of affairs that emerged as recently as 5,000 years ago, when the first complex societies started to appear. That's not to say that all human groups before this time were the same: even the earliest humans are likely to have lived in groups that differed more from one another than, say, the average meerkat or babbler or chimpanzee group does today. Nevertheless, the earliest human societies are thought to have shared some fairly important features. For most of our time on Earth, we lived in low-density societies, with no fixed abode and relied on food we could either hunt or find (rather than grow or buy). Our societies differed from those of the other apes in several important respects, many of which we've already covered: we were more cooperative, our group boundaries were more flexible, and we had families. But that's not all. The earliest human societies differed from those of chimpanzees or gorillas or bonobos in another crucial aspect: they were egalitarian.

Classifying human societies as egalitarian might come as a bit of a surprise, particularly if we think of equality purely in terms of money and possessions. But money and the wealth it creates is a relatively recent human invention, and our earliest ancestors would not have had houses to call their own, or accumulated very many possessions at all. A more relevant evolutionary currency for measuring societal equality is reproductive success. For most of our time on Earth, by this measure, our societies would have been relatively fair: most individuals would have had similar numbers of offspring. Contrast this to the situation in hierarchical great-ape societies, where alpha males are able to ring-fence resources and monopolise females to the exclusion of lower-ranking individuals in the group. In some species, such as gorillas, reproduction is a winner-takes-all scenario, with the alpha male siring nearly all the offspring in the group. In chimpanzees, low-ranking males do breed but are far less successful than the alpha male, who can expect to sire a third of the group's offspring and more than twice as many as his closest competitor. Although we do not have data for reproductive equality in ancestral human populations, we can use measures taken from contemporary hunter-gatherer societies to make some informed estimates. Large-scale studies that have examined the link between a male's status and his reproductive success in such societies find that status matters, but not nearly as much as in more hierarchical ape species. Reproduction tends to be shared far more equally among human males than it is among any of our closest living relatives.

Alpha-male chimps get what they want by throwing their weight around, and physically intimidating or bullying their rivals. But, in ancestral human groups (as well as in many contemporary foraging societies), this archetypal great-ape route to the top was effectively sealed off. In many foraging societies, big-shot behaviour is more likely to result in ridicule and ostracism than in power and glory. Personal autonomy is prized above all else, with the ethos of group life being that people can pretty much do what they like so long as this doesn't involve harming or bullying others. Or, as one Ju/'hoansi man put it when asked by an anthropologist why they had no headmen, 'Of course we have headmen. Every man is headman of himself.'

In egalitarian societies, the route to high status goes via prestige and respect, rather than force and intimidation. It is the path of persuasion, rather than power. A high-status individual is typically respected, generous and the kind of person who acts in the best interests of the group. To maintain good standing, individuals must be seen to shrug this status off and avoid acting in a manner that invites accusations of self-aggrandising or coercive behaviour. People who ignore this tacit directive can expect to feel the ire of their group mates. Among the Ju/'hoansi, a man who is getting too big for his boots or bossing others around might be addressed as 'Big Chief': a sardonic rebuke that carries the threat of further sanctions if unheeded. Countless anthropological reports detail how men who throw their weight around, or try to monopolise women or other resources in the group, can expect to be ostracised, excluded from the group, or even killed. The short passage overleaf describes the methods the Kalina (a Guyanan tribal society) use to deal with bullies and big shots. Such methods are widespread across non-industrial societies and were probably also characteristic of earlier human groups more generally.

> The men of the settlement will talk to him, but if he does nothing to improve his position in their eyes, he will be advised to leave on pain of having life made very unpleasant for him. If he persists in remaining he will find that he and his family are social outcasts: they are not invited to drinking parties; he will be unable to borrow anything; he will get no help in hunting, fishing, field cutting, canoe building or other activities in which the men assist one another, nor will his wife receive aid in her occupations; his household will be excluded from the waterhole and bathing place. In short, he will lose all advantages of group life. In aggravated cases, the other men may beat him or even kill him, if he fails to take the hint.

Importantly, the lack of coercive dominance in ancestral humans and modern forager societies does not stem from an intrinsic aversion to hierarchy: the drive to compete with others for mates, food and other resources is a general feature of most *animal* brains, and all of the great apes. We are no different.

Instead, I think it makes the most sense to conceive of the distribution of power in human societies as a giant tug of war. On one end of the rope is the individual urge – present in all of us to varying degrees – to predominate over others; to use any talents, skills or abilities, or any fortuitous circumstance, to rise just a little bit higher than the rest. Pulling on the other end of the rope is a force we can sum up as the collective interests of everyone else. As with any tug of war, often one side prevails over the other. The collective-interests team may seem to have greater strength in some societies or to have prevailed in some historical periods, but there are also times where 'individual-interests' seem to pull harder, with small cliques of elites – tyrannical monarchs, emperors and dictators – managing to subdue and aggressively dominate much larger populations.

This tug-of-war analogy illustrates two important principles. The first is that humans are not intrinsically opposed to seeking or being in power; in fact, most humans care about status and wealth and, in some cases, benefit hugely from the opportunities that social dominance provides. The second principle is that an absence of power within societies is not an omission, or a missed opportunity, or something that only exists because no one thought to take it. Instead, a power vacuum is the outcome of a constant tension, something that is actively maintained through the efforts of the many against the few.

*

If you're feeling a sense of déjà vu, it's because we've met these socialist principles before. We saw earlier how parliaments of genes collude to prevent selfish elements from biasing their appearance in the gametes; and how workers in a social insect colony police one another, preventing their sisters from laying furtive eggs. Now we see these mechanisms at play again, this time in the transition from what were ancestrally despotic to egalitarian human societies.

To understand why individuals might put aside their own personal ambitions in pursuit of a collective goal of equality, it helps to consider the stakes in these kinds of competitions. We can safely assume that every gene and every individual will try to prosper – but there are two potential routes to victory. One is the high-risk, winner-takes-all competition, where everyone battles it out against everyone else in the pursuit of individual glory. This is reminiscent of the analogy I used in Chapter 3, likening selfish genetic elements to queue-jumpers. If all genes (or individuals) act in this way, the outcome is a free-for-all, where everyone tries to elbow their way to the front of the line. But there can only be one winner: the person who ends up at the front. If you win a competition like this you win big but the most likely outcome is failure: if you end up in last, third or even second place in the queue, you get nothing.

The alternative strategy is to join a team, much like many people form syndicates rather than playing the lottery individually. Being part of a syndicate means that you have a better chance of winning something in every lottery draw, even if you have to share that prize with your teammates. Being part of a syndicate is therefore a more prudent strategy: it sacrifices the small chance of achieving solo victory in order to minimise the chance of failing to win anything. And, over evolutionary time, the more prudent strategy tends to succeed.

*

Given our abilities to hold dominant, coercive individuals in check, you might be wondering how tyrants and bullies ever manage to prevail. Over the course of our short time on Earth, we seem to have swung wildly from one extreme to the other, from egalitarian forager societies to the intensely despotic regimes that emerged at the dawn of civilisation. How did we end up with emperors and monarchs, and the Saddam Husseins and Adolf Hitlers of the world? And why did the humans living in the earliest despotic states accept being coerced by others, when their ancestors had rejected this way of life?

To sketch out how this might have happened, we need to go back to the period when humans first started to settle down, around 12,000 years ago. This period is sometimes known as the Agricultural Revolution because it is the period when humans began farming crops and keeping herds of domesticated livestock. The transition from a nomadic way of life to a more sedentary existence was permitted by a period of climate change, with the prevailing conditions becoming vastly more stable than they had been in the millennia before. Climate stability would have been conducive to farming because crops were less likely to be wiped out by environmental shocks, and because it allowed the earliest human farmers an opportunity to experiment with different farming practices, identifying the methods and species that were most amenable to mass production, without having to also contend with wild fluctuations in weather.

Success in these early ventures, like all collective action, relied on coordination; and this is where leaders come in. We've already seen that leaders exist even in otherwise egalitarian hunter-gatherer societies, but only in specific contexts, such as when there is a need to coordinate action during hunting or to mediate a dispute among camp members. In these societies, any authority a leader wields is bound to the context in which it is needed, meaning that leaders in one domain typically lack authority in any other. For example, among the Chabu of Ethiopia, a man who owns a pack of hunting dogs might decide when a hunt will take place and where the hunters will go. He will orchestrate tactics during the hunt itself and, if the hunters are lucky enough to catch something, the dog owner will also be responsible for dividing up the meat from the kill. Although this man leads the hunting expedition, his authority evaporates when it comes to other realms; he is leader of the hunt and the hunt alone. Living in an agricultural society would have been different because, unlike a hunting mission, farming is not a finite activity, something which has a definitive start and end time. Tending to crops and animals is a way of life, allowing leaders to become a more permanent feature of society.

The earliest forms of institutionalised leadership were likely to have been inclusive and democratic, with the leaders being those who were best able to build consensus and coordinate actions. These early leaders probably therefore had much in common with those who were (and are) afforded high status in mobile foraging groups: they were people who commanded respect and were well liked and tolerated by other members of the group. Despite starting out as largely egalitarian, however, these early human civilisations somehow morphed into despotic regimes, where leaders become predatory and domineering, subjugating the many serfs under their rule while creaming off the benefits of the surplus generated by this workforce. This obviously didn't happen overnight – it would have been a gradual process occurring over generations, with followers giving up increasingly large shares of resources and coercive control to leaders.

To understand why the followers might have accepted this (seemingly terrible) deal, let's look at a theoretical model, published in 2014, which describes in mathematical terms how a hypothetical 'acceptance of hierarchy' trait could spread in a population where most people start with a preference for equality. Don't worry: I'm not about to fill the page with mathematical equations as the main conclusions can easily be grasped with just a verbal description. Theoretical models are especially helpful when it comes to understanding human evolution because we can't go back in time to measure the reproductive and survival consequences of different modes of life. Models act as digital Petri dishes, allowing us to specify the starting conditions for the hypothetical world we are interested in and then see how things might have played out. Nevertheless, it is worth bearing in mind when evaluating this (or any other model) that the results hinge fundamentally on the starting conditions – or assumptions – we make about the way the world is. Or, to put it another way, if you put rubbish in, you'll get rubbish out.

In this case, the model is framed in terms of understanding a tolerance for hierarchy, but you can also think of it as a model for exploring this transition in lifestyle, from living in an egalitarian foraging society to living in a hierarchical farming society. This approach allows us to ask how and why we adopted the farming way of life – and the inequality that came with it – when we seemed to have a pretty good deal living in more egalitarian, foraging societies. The model uses reproductive success as the ecological currency, specifically asking whether individuals who prefer life in hierarchical farming societies tend to have more offspring than their counterparts living in egalitarian forager societies. In this way, we can link psychological traits with the outcome that evolution cares about: differential reproductive success.

The model is built on a handful of uncontroversial assumptions. First, we assume that leadership facilitates coordinated action, meaning that farming societies are more productive than foraging societies. Second, we assume that any surpluses that are generated through collective action (hunted meat, or crops produced, for example) are shared more equitably in the foraging societies than in the farming societies. At this point, we aren't invoking all-out despotism, but just the more plausible scenario where the leader of a farming group keeps slightly more of the resource for himself. The final assumption is that these two kinds of groups coexist in space. This means that an individual's reproductive success is measured relative not just to others in his own group, but to the population at large. This part is important, because it means that getting a smaller share of a large pie (the farming life) can be better in absolute terms than getting an equal share of a small pie (the foraging life). Because resources translate into reproductive success, people living in farming societies can have more offspring than those who live in the foraging societies. In the language of the model, this means that selection could have favoured a tolerance for hierarchy, even if this went hand in hand with a degree of inequality.

Although we can understand how some degree of inequality may have been tolerated in the first sedentary societies, it is harder to fathom why scores of subordinated individuals put up with the extreme despotism that we see, and have seen, over the course of human history. The consequences of such extreme inequality can most easily be appreciated by looking at disparities in male reproductive success in ancestral foraging societies compared to those of men living in the first known civilisations. Estimates based on the period before humans migrated out of Africa (around 100,000–200,000 years ago) suggest that there were around three reproductively active females for every reproductively active male in these ancestral human populations. That implies that for every male who sired offspring, there were two others who did not. By the middle of the Holocene, once the Agricultural Revolution was well under way, this ratio increased to around sixteen, implying a dramatic increase in the number of childless men. Although hierarchy would not have been the only cause of this reproductive skew (men were also more likely to die than women, for example during warfare), status also played an important role in determining who sired children. In these earliest civilisations, some of the most powerful emperors and kings fathered hundreds of children, while others had none. Many men were castrated in childhood, forced to work as eunuchs within the leaders' palaces. The ritualised killing of other humans became common, both as a means to demonstrate the great power of the authorities and as sacrifices to deities. And with the rise of hierarchy, we also witnessed the advent of the most oppressive kind of subordination: slavery.

How could life under extreme despotism ever be preferable to life in a foraging society?! The answer is that it wasn't. People didn't choose this way of life; they got stuck. To see why, let's go back to our digital Petri dishes, where we can watch how this next period of societal evolution might have played out. We know that the early farming groups were more productive than their neighbouring foraging groups, and that people may have tolerated a small degree of inequality in return for increased reproductive success. But populations are seldom static. In our digital Petri dish, the farming groups start to expand, a simple consequence of the group members' increased reproductive success. As these groups expand, they start to encroach into the available space, pushing the foraging groups closer and closer to the edge, until the latter disappear from the population entirely. This is an important point of no return, because now the subordinates living in the farming societies have no alternative lifestyle to revert to. Leaders who previously took a slightly larger share of the pie could now take the majority of it, with the subordinates having to accept the deal because they were now locked into this way of life.

This model is abstract, and yet it reflects what we know about the rise of inequality over time. The prediction that hierarchy should be most intense when followers cannot leave is supported by historical accounts. The creation of ancient Egyptian states, where godlike pharaohs presided over farmers and slaves, was facilitated by the geographic location of these kingdoms, along the banks of the River Nile. Surrounded on all sides by harsh desert, it would have been prohibitively costly for dissatisfied minions to up sticks and walk away. Similarly, in Peru, early states emerged in the narrow fertile valleys where agriculture was practised, which made it difficult for citizens of these early states to leave. Slavery was also historically common in sedentary tribes, such as the Kwakiutl and the Chinook of the Pacific Northwest coast, showing that even where farming doesn't exist, transitions to despotism are more likely where the subjugated members of society cannot easily leave and join other groups.

*

Being surrounded by desert or other inhospitable territory can make it difficult or even impossible to walk away from a repressive regime, but subordinates could have rebelled by rising up and overthrowing an oppressive regime. Despite being possible, such events are likely to have been infrequent, an observation that is especially puzzling given that the individuals in the subordinated class would have far outnumbered the higher-ranking elites. In a duel between a handful of tyrannical alphas, and a legion of united subordinates, the wise money should be on the side with the numerical advantage every single time. This principle of there being power in numbers is so universally predictive of success in combat that it has been summed up as a mathematical proof, known as Lanchester's Square Law, a rule so pervasive that even preverbal infants intuitively understand it. The rule states that the fighting strength of any group is proportional not to the absolute member number, but to its square. That means in a combat of two against four, the fighting strength becomes four against sixteen. When we start scaling things up to nation-state level – a few hundred political elites against a population of millions – we can easily appreciate the formidable asymmetries in the raw power of the two groups.

So, why don't the populations of dictator-led regimes use their numerical advantage to rise up against their tyrannical leaders? Why don't the millions of people who vote for policies that are not enacted, or who disagree with their leaders, storm governmental offices and really 'take back control'? The answer is that revolution is a collective-action problem, which is, of course, a problem of cooperation. A collective is powerful when everyone acts in unison, but insurgence entails risks. Each individual therefore faces a temptation to stand aside and let others pay the costs of insurrection, while enjoying the benefits that any success might bring.

As a particularly chilling example, take the vanishingly small frequency of rebellion aboard the ships carrying hundreds of Africans to a life of servitude across the Atlantic during the slave trade. Considering the conditions aboard the slave ships – and the terrible life that awaited the captive men and women who managed to survive the voyage – it is surprising that as few as two in every one hundred ships experienced a rebellion, where the slaves fought back against their captors. Staging an uprising would have been practically difficult, of course: captives were locked below deck and chained in pairs – often to a complete stranger, who might have hailed from a different region, and who spoke a different language. Preventing communication between the captives was no accident: captains conspired to populate the ships with people from different ethnic groups, knowing that the language barrier would severely hinder any attempts to engineer a rebellion.

But iron shackles and language barriers weren't the only reason that the captives so infrequently rebelled against their captors. Instead, as a 2014 study suggested, the decision to sit tight may also have been a judicious one, based on the likely costs and benefits of participating in a joint mission to secure freedom. The costs of being implicated in a conspiracy to revolt were severe, and captains would mete out barbarous punishments to rebels, to deter others from following suit. Organising a rebellion required slaves to trust one another, so captains sowed distrust and suspicion among the captives by offering freedom to any individual who spied on the other prisoners, and informed the captors of budding plots aboard the vessel. This further raised the risk of being the first-mover in a collective-action problem: coordinating a rebellion would be even more fraught when you couldn't be sure if your co-conspirator was secretly planning to turn you in.

The balance of these incentives resulted in a rather counter-intuitive outcome: although we know that power increases with numbers, increasing the numbers of slaves aboard a ship served to quell, rather than stoke, an insurgency. Adding a hundred captives to a ship, for example, is estimated to have reduced the risk of rebellion by up to 80%. To understand why, it helps to think again about the likely costs and benefits of taking part in dangerous collective action. In large numbers, each individual's contribution to a revolt will make only a small difference to the chance of success. Meanwhile, the costs of being discovered as a conspirator remain just as stark. With such odds, it might be better to wait for others to make the first, most dangerous move, and hope to benefit from the freedom that a successful rebellion could bring. In a smaller group, the incentive to contribute becomes higher, because each individual plays a more pivotal role in the success or failure of the mission. In smaller groups, it is also easier for individuals to monitor the contributions of others, and to persuade or shame each member into participating.

The implications for the vast majority of Earth's population, where an ever-diminishing minority of individuals and organisations are profiting from the wholesale destruction and exploitation of our planet, are stark indeed. In these global social dilemmas, the stakes do not seem to be so high as for the men and women aboard the slave ships, but the incentives are broadly the same. Collective action in the face of climate change requires commitments from vast numbers of people. This means there is a risk: if you make a personal sacrifice by deciding not to fly or not to have another child or to eat a plant-based diet, there is a reasonable chance that this sacrifice will be for nothing. Most of us would benefit if we could find ways to take back control of this situation. But to do so requires collective action on a scale we have never seen before.

18
VICTIMS OF COOPERATION

Every sin is the result of a collaboration.

Stephen Crane, 1898

Each night, as the planes land at Reagan National Airport and the passengers spill out of the doors, the throng of Uber drivers waiting outside collectively switch off the app that indicates their availability. Then they wait. As the minutes pass the price creeps up, demand adjusting to supply in real time. $12. $13. $15. Eventually, the ringleader gives the signal and in unison the taxi drivers switch the app back on, accepting fares from passengers who will pay a little bit extra to be taken to their destination.

Artificially squeezing supply to generate surge pricing might not sound like a particularly cooperative thing to do – quite the opposite! But cooperation and competition are simply two sides of the same coin. What looks like cooperation through one lens will often be felt as competition through another. The drivers who switch off their app may be exploiting their passengers, but they must work together to do so. Although the best collective outcome is that everyone turns off their app and generates the surge pricing, there is also a possibility that there are more drivers than passengers seeking rides. Each individual driver therefore faces a selfish incentive to free-ride, to turn his own app back on a bit sooner than the others to ensure that he doesn't risk missing out on a fare altogether.

This story helps us to confront a possibly uncomfortable truth, one that has been bubbling away in the background of all that we've encountered so far. Cooperation is, at heart, a means by which entities improve their own position in the world. In other words, cooperation is favoured if and when it offers a better way to compete. A corollary of this is that cooperation frequently has victims (in fact, cooperation *without* victims is the most difficult kind to achieve).

By viewing cooperation in these terms, we can recast phenomena such as corruption, bribery and nepotism as forms of cooperation where the benefits fall to one's kin or collaborators and the costs are felt by others in society. Preferentially hiring a family member for a job or greasing the palm of an executive to secure a lucrative contract are both cooperative interactions, involving help and trust. If these activities strike us as nefarious it is because this local cooperation frequently produces societal costs. Cancer is a useful example to recall here. In Chapter 3, we saw that cancer cells form cooperative partnerships and that working together is a means by which these alliances are able to outcompete our other cells. But cooperation among cancer cells results in severe and potentially lethal consequences for the host. Whether we view cooperation as beneficial or harmful therefore depends entirely on the perspective we adopt.

*

If someone were to ask you whether it would be acceptable to lie in court to exonerate a family member, what would you say? What about if you were asked whether you had a moral duty to hire the best candidate for the job rather than a less-qualified friend? Answers to these sorts of questions are neither straightforward nor universally endorsed, because cooperating at one scale is often traded off against cooperation at another. Our sense of what is moral or immoral depends on how we feel these competing interests ought to be balanced. To put this another way, one might mistrust someone who 'always helps his friends'. But a similarly damning accusation can be levelled at someone who 'doesn't even help their friends'.

In what follows, it will be helpful to think of yourself as existing inside a circle, with all the other people on this planet existing somewhere inside that circle with you. The people who stand closer to us – our family and close friends – are those to whom we feel the greatest moral obligations, and for whom we show the greatest concern. This much is unsurprising: we've already seen that cooperation and interdependence are higher among related individuals and in enduring relationships. We have a stronger stake in the well-being and success of these core members of our social circle and it makes sense that our prosocial efforts are more likely to be bestowed upon them.

But other people exist inside this circle with us, albeit standing slightly further away. These more distant connections are not treated equivalently to our family and friends but they are nevertheless people whose fortunes do concern us, and to whom we feel we ought to extend a degree of help and trust, should the situation demand it. Some of the largest differences in how people see the world – and the source of some of the most profound disagreements humans have with one another – lie in determining just how much we ought to prioritise our nearest and dearest over these other connections.

*

In the interest of outlining a clear story, I'm going to speak in general terms here, but please don't mistake this for an assertion that we can pigeon-hole people as being morally parochial or morally universal just from the colour of their political stripes,* or based on the country or region they were born or live in. As an analogy, we know that men are on average taller than women, but even if you knew someone's height you wouldn't bet the farm on guessing their sex. The reverse is also true: knowing a person is either a male or a female doesn't let you make precise predictions about their height. Interpret the patterns I will describe here in the same way: they are broad trends and cannot be used to make assertions about any particular individual. But these broad trends do have real-world implications for the functioning and structure of the societies we live in, and the kinds of relationships we have within them.

The extent to which people feel that cooperation should be kept within a smaller social circle, rather than extended to more distant connections varies to some extent along the political spectrum. This is nicely illustrated in the inaugural addresses given by US presidents in recent years. In 2009, Barack Obama declared that 'America is a friend of each nation, and every man, woman and child', and promised to work alongside the 'people of poor nations ... to nourish starved bodies and feed hungry minds'. In 2017, Donald Trump drew a much smaller circle around the interests of the nation rather than the planet, bemoaning the 'trillions of dollars [spent] overseas' and promising that 'From this moment on, it's going to be America First.'

Large-scale behavioural experiments have confirmed these patterns, revealing that people who describe themselves as being more politically conservative profess greater love for their family, but less love for humanity as a whole (with political liberals showing the opposite pattern). In the same study, political conservatives rated protection of their nation from enemies as being more

* Indeed, arguably the most popular US president of all time in Africa was a Republican – George W. Bush. During his presidency, Bush established PEPFAR (the President's Emergency Plan for AIDS Relief), spending $80 billion (by 2009) and saving the lives of an estimated 13 million people living with HIV or AIDS in Africa.

important than establishing a world free of war and conflict; and political conservatives also tended to identify more strongly with their immediate community than with humanity as a whole. When asked to allocate hypothetical resources across different categories – to family and friends, to all humans and even to non-human animals – political conservatives tended to allocate these resources more narrowly than liberals and were less likely to allocate anything to non-human species. In a recent study I did with Lee de Wit (a researcher at the University of Cambridge), we found that people's concern about the impacts of COVID-19 also varied along the political spectrum. While people of all political stripes were worried for themselves, and for their close friends and family, broader concern about the impacts for other people in society was more keenly felt by political liberals. In sum, these experiments reveal a smaller circle of moral regard among political conservatives, meaning that empathy, compassion and concern are allocated in greater quantities to more proximal connections than to more distant ones.

*

These circles of moral regard – that dictate the extent to which cooperation should be preferentially extended to closer connections versus shared more equally among everyone – also vary on a much broader scale, across countries. These cross-cultural patterns are sometimes described in terms of differences along a universalist–collectivist spectrum. Collectivist societies (such as in China, Japan and Korea) tend to be built around family groups. In such societies, social circles tend to be relatively small, but the links within them are extremely strong: individuals greatly depend upon one another to get by. People have strong moral obligations to help those within this inner circle, but need not extend such favours to those outside this core group. At the other end of the spectrum, in universalist societies (like in many countries in Western Europe, and the US), people tend to have larger social networks that include many distant connections, but where the ties of moral obligation to close family are correspondingly weaker. Although people still preferentially help and trust friends and family, there is not the same moral imperative to help this core group to get ahead. Instead, moral norms in universalist societies emphasise impartial prosociality, meaning that the same rules should apply to everyone.

The size of these social circles can account for some of the large-scale differences in how societies function. For instance, collectivist societies tend to experience higher levels of corruption, bribery and nepotism, all of which can be understood as prioritising the needs of those *inside* the circle of moral regard over the needs of those on the outside. Appointing friends and family to executive roles (rather than making meritocratic hires) is more common in cultures with stronger family ties, and collectivism also predicts a stronger tendency to endorse breaking the law, for example by lying in court, if doing so will help a friend.

As you might expect, collectivism (or strong family ties) is also associated with a reduced trust in strangers, which can be measured both through surveys and in real-world behaviours. A particularly illustrative case is Italy, where family ties are stronger in the south than the north.* Italians who hail from southern regions trust less in institutions, keeping a larger proportion of any household wealth as cash rather than invested in banks or in shares. When taking loans, Italians from southern regions are more likely to borrow from friends and family than from banks; and transactions are also

* Some researchers have argued that the church may have had a hand in producing these regional differences. From as early as AD 500, the church banned marriages among cousins and extended kin. This meant that young people, upon reaching marriageable age, frequently needed to relocate in order to find a husband or wife whom they were allowed to marry. These enforced marriage patterns both physically and genetically diluted extended family networks, and are thought to have given rise to the rather unusual isolated, nuclear families that many Westerners live in today. In Italy, the church instantiated the ban on kin marriage in the north earlier than it did in the south, a fact that might explain contemporary cultural differences between these regions.

more likely to take place using cash, rather than cheques or forms of credit. Collectivism also predicts a reduction in the tendency to help strangers: blood donations are lower in the south than in the north of Italy and a recent experiment employing a 'lost letter' design (where stamped, addressed letters are left on the street and the experimenter measures how many are posted) found that letter return rates were higher in the north than in the south. The general pattern here is that strong family ties increase cooperation and trust inside the immediate social circle, but decrease trust and cooperation beyond this boundary.

These kinds of effects can also be observed in large, multi-country studies. In one colossal experiment conducted in 2019, a team of scientists dropped more than 17,000 wallets over 350 cities around the world and explored the factors predicting whether the wallets (which contained money, and a name and address) were returned by members of the public. Returning a wallet containing money to someone you have never met (and will probably never meet in future) is a reasonably robust measure of willingness to help a stranger. One of the key findings was that the wallets were more likely to be returned in 'universalist' countries compared to when they were dropped in countries where people have stronger kin ties.

*

We should resist interpreting such findings with a moral overtone: trusting in and cooperating with kin, or inside a small social circle, is not *necessarily* worse than trusting and cooperating with those beyond the kin group. Quite the opposite: if this is how others in your society are likely to behave, then focussing your cooperative efforts on kin and close friends is an eminently rational strategy.

Another way to quash the moral implications of these findings is to query the foundations of these differences in the scope of moral regard, to ask where they come from. To do this, let's start with three ecological variables that have concerned our species since the dawn of our time: threats, sustenance and disease. These three concerns are things that *really* matter. If we can avoid being attacked or harmed, and we can get the food we need and stay healthy, our most basic needs have been met; this is the essence of what's called 'material security'. To achieve it relies fundamentally on cooperation. Cooperation is therefore a form of social insurance: a way of buffering the risks of not meeting one or all of these basic needs in life.

For most of our time on Earth, this insurance has come in the form of close social networks, comprising friends and family. For many people, these local, individuated relationships are *still* the primary means to buffer life's risks. In many non-industrialised societies, people routinely share food with neighbours and friends. Food sharing is a means to dampen the peaks and troughs that would otherwise ensue when people don't have access to external market-based exchange. You might also remember the *osotua* relationships of the Maasai herders, which allow the risk of losing cattle to be pooled across a bonded pair, with each partner committing to help the other should the need arise. Pooling risk across a few highly interdependent relationships is the primary means by which humans managed to survive, and thrive, in the harsh and unpredictable environments in which we evolved, and for many humans such relationships remain the primary form of social insurance to this day.

But for those of us living in modern, industrialised societies, things look different: the state has largely taken the place of these interdependent relationships, and provides the infrastructure and support to ensure our basic needs are met. By providing public services, such as armies and healthcare, the state protects us from existential threats and disease. By enforcing rules and norms of trade, the state allows market economies to flourish and for resource surpluses to be generated. A state-backed currency allows us to store this surplus, as money in banks; and this stored wealth allows us to buffer our own supply chain, meaning that we can reliably gain access to the resources that we need without having to rely on help from others.

Material security fundamentally alters the shape and size of the social worlds we inhabit. Low material security tends to go hand in hand with small social networks: when we need to ask more of one another, we nurture a small number of highly dependable relationships. As material security increases, this weakens people's reliance on close, interdependent relationships – and their investment in these relationships typically dwindles as a consequence. When material security is higher, people can also afford to expand their networks a little, seeking out the opportunities that come from establishing new partnerships where the stakes are not so high.

This highlights the fundamental role that the state can play in shaping the social worlds we live in. If the state will ensure that our most basic social needs are met then we no longer have to rely on a few, highly interdependent relationships to provide material security. Freed from the existential threat of not meeting our basic needs, we can also afford to take a few social risks, and the boundaries of our social circles can relax a little, expanding to include people from beyond the core network of family and close friends. The state can further support these interactions beyond the core group by enforcing rules that constrain individuals' abilities to swindle one another, and (for the most part) promote mutually beneficial exchange. By providing a safety net for our basic needs, and a set of rules to facilitate mutual exchange, a functioning state allows individuals to draw larger circles of moral regard around themselves and to endorse universalist, impartial norms of cooperation. Functioning states – and the institutions they embody – are the foundations upon which modern democracies are built.

*

Material security varies hugely, both within and between countries. Some of this is due to basic geographical factors: for example, pathogen prevalence varies with latitude, being highest at the equator and diminishing as we move further from it. This means that people living closer to the equator will tend to experience a higher threat of pathogen-borne illness, and a corresponding decrease in material security. Other natural threats, such as extreme weather events and food availability, also vary along this dimension, further compounding the issue. Even where geography is held constant, we might observe differences in material security due to personal circumstances: wealthy individuals can store money in the bank to ensure a steady supply of food, but many people do not have the means to buffer their supply chain in this way.

For the most part, variation in material security is relatively stable, in that we can predict it from one year to the next. In the US, the Southern states are always going to have higher risk of extreme weather events; and a person who is on the breadline one year is (unfortunately) most likely to remain there in the year to come. But every now and then we experience a shock to material security that comes out of nowhere, something that can knock entire nations off balance and calls our very existence into question. Such an event happened in 2020, when a deadly coronavirus emerged from Wuhan in China, spilling across country borders, and infecting nearly half a million people around the world within a matter of weeks.

Although the mortality rate of COVID-19 is relatively low compared with other recent epidemics (e.g. Ebola, SARS*), the sheer scale of infection risked overwhelming hospitals. This mattered, not just for people who fell sick with the virus, but for anyone needing

* At the time of writing, the WHO's estimated case fatality rate for COVID-19 (the percentage of people known to have been infected who have a fatal outcome) is 0.6%. This is substantially lower than the case fatality rates for Ebola (which can be as high as 90%), SARS (around 15%) and MERS (around 35%).

urgent and unexpected hospitalisation: if a system is at breaking point, then its ability to provide care to anyone who needs it will be severely compromised.

In this situation, our typical means for achieving material security became less effective. Faced with the prospect of being quarantined at home with the disease, people began to stockpile groceries to survive the confinement. A once humorous observation that people were panic-buying toilet roll became a source of genuine anxiety: what would we do when we needed some? Glib assurances from government officials that there was 'enough food for everyone' were readily contradicted by our own experience: supermarket shelves were stripped bare; food delivery services were overwhelmed; and getting into the shops involved standing in a queue for hours on end. As the number of serious cases threatened to swamp hospitals' intensive care capacity, the state provision of healthcare was called into question. If we fell ill, there was no longer a guarantee that we would be adequately cared for.

What we saw in the wake of this crisis, however, was a resurgence of cooperation. Local social networks sprang up, seedlings of cooperation germinated by our shared fate and dependence upon one another. People posted letters through neighbours' doors, providing phone numbers and promises of help if needed. Neighbours shared food once more. Local shops rallied round to provide for the most vulnerable members of the community, and many offered free deliveries to those who, for health reasons, could no longer mingle in public spaces. A bookstore in my neighbourhood closed its doors to customers, but provided book sales over the phone. When I called to buy a couple for my youngest son (who was unfortunate enough to have his birthday fall in late March), the owner gift-wrapped them and, sensing that I needed something to take my mind off things, threw in her favourite novel for free. In the face of extreme hardship and uncertainty, we reach out to one another, offering what little we have rather than turning to face the other way.

These stories of cooperation are heart-warming. But they are also circumscribed, occurring at a hyper-local scale. In the face of threat, our scope of moral regard tends to shrink: we may readily help our neighbours, but continue to defect on a wider scale by stockpiling groceries at the store, leaving less for others. In a crisis, other shoppers from different neighbourhoods were more likely to fall outside our shrunken scope of moral regard.

Similar patterns also emerged on a national scale. In the US, the value of being 'United' rather than just 'States' – part of a collective rather than fifty separate entities – is that different states can help one another out in times of need. A state struck by a fire or a tornado can rely on others for help, an arrangement that works because these shocks typically don't hit every state at once. The spread of a pandemic virus changed this situation dramatically, as states were either experiencing or preparing for the inevitable emergency. And interstate cooperation dwindled as a result: no state could afford to help another to secure ventilators or protective kit for medics or testing kits if this meant that they might end up with a shortfall. At an international scale, European countries were reluctant to help neighbours who were overwhelmed by the disease, while the US president Donald Trump was criticised for buying up the world's stocks of remdesivir, a drug with the potential to help patients recover from COVID-19. As vaccines become available, we might expect to see similarly parochial interests surfacing as nations scramble to secure access for their own citizens first.

*

In the face of global problems, we need global cooperation. And pandemics are not the only, or even the gravest, problem we face. The list is seemingly endless. Anthropogenic climate change, habitat loss, species extinctions, rising levels of pollution, over-consumption of finite resources and the inability to commit to nuclear disarmament are all up there in the long and depressing list of ways that we – as humans – are failing to cooperate to achieve a greater good. More than 30% of the world's fisheries have been overfished to the point where they now face the risk of collapse. Some species, like bluefin tuna, are likely to go extinct in our lifetimes. During the summer of 2019, more than 30,000 fires raged in the Amazon, many of which are reported to have been deliberately caused by famers and loggers to clear land for crops or livestock. The world's insatiable demand for palm oil products means that 80% of orangutan forests will be decimated by 2080, and this iconic ape – this cousin of ours – will join the long list of species that no longer exist because of us.

Solving problems like this is difficult because doing so requires us to cooperate on a global scale. Moreover, global public goods can be enjoyed by everyone, even those that don't contribute to providing them. If we managed to reduce air pollution in the centre of London, we couldn't prevent SUV drivers from breathing it. If we could prevent greenhouse gas emissions from exceeding a dangerous threshold, we couldn't stop frequent flyers from enjoying it. In a social dilemma, individuals can benefit by free-riding on the investments of others, while still enjoying any benefits that collective action brings. The free-rider problem underlines the difficulty in finding cooperative solutions to global public goods problems, even though our tenure on this planet depends so crucially upon us doing so. When we play a global game with everyone on the planet, cooperation pays less than selfishness because cooperation offers no way to get ahead, no scope for relative advantage. Our failure to cooperate may spell disaster in the long term, but this timeline is far beyond the horizon that we typically consider. This stark evolutionary logic seems to have us racing towards a cliff edge, cognisant of our fate, and yet seemingly powerless to stop and call a truce.

The benefit from pursuing a short-sighted, self-interested approach can thwart the emergence of global cooperation. For example, in January 2019, a rare bluefin tuna weighing more than

270 kilos sold for $3.1 million (333.6 million yen) in Tokyo's Toyosu fish market. That's a lot of money to turn your back on, especially if you believe that if you don't catch the fish and sell it, someone else will. On a much smaller scale, we all face variants of these dilemmas on a daily basis. Why should I avoid flying, when no one else will? I really want these new jeans – do I need to think about the environmental cost of producing them? This sorry state of affairs can result in what's known as the 'tragedy of the commons'. We need to resolve it, but what will it take?

From everything we've seen, we know that it is both naïve and dangerous to rely on lukewarm appeals to our better natures to tackle problems on this scale. For example, Big Society – a political ideology coined in 2010 by the Conservative Party – aimed to transfer power for producing public goods from politicians to the people, one of the main tools in the toolkit being to encourage volunteerism. We now know that this failed spectacularly – and this book has given some insights as to why. As we have seen, we aren't unconditional cooperators. Evolution would never have favoured such an indiscriminate willingness to help. Rather, we are more circumspect, we fine-tune our investments to the situation and to the possibility of downstream benefits in the future. If the incentives don't favour it, we simply don't cooperate all that much.

To illustrate the difficulty in engineering cooperation, let's use 2020's coronavirus pandemic as a sobering example. By now it seems clear that the mortality rate is relatively low among young, healthy people and much higher in people with underlying health issues and the elderly. As with any infectious disease, mitigation strategies rely on cooperation; people's willingness to curtail their normal activities, their routines and travel schedules. If the majority of people would adhere to these measures, we could reduce the costs which would be borne by the most vulnerable members of our society. But cooperation can be difficult to achieve because we face a collective-action problem. Why should any individual pay the costs of changing her routine if others aren't doing the same? Why should panicking shoppers refrain from stockpiling toilet rolls and pasta, if they believe that they might otherwise be left without? Why should people wear face masks to protect others or adhere to quarantine measures, if they see or believe that others are flouting these guidelines? In the UK, early advice and pleas from government officials went unheeded; and it was inevitable that, after a few weeks, we were no longer asked but told. Stricter measures were introduced, with the threat of penalties for non-compliance. These measures were needed and probably should have come sooner: relying on people's goodwill is not just deeply naïve; in the case of a global pandemic, it costs lives.

Having watched responses to this pandemic play out on a global scale, it is difficult to remain optimistic about our ability to tackle the much larger problem we face, of anthropogenic climate change. Pandemics have many features that render them intrinsically easier to solve: the threat is here and now; citizens don't want to catch the disease if they can help it, and so are motivated to take strong action to protect themselves; and there is a strong economic incentive to eradicate the disease as quickly as possible. In the case of the 2020 coronavirus pandemic, some of the later-hit countries could observe how the virus spread elsewhere, and thus had an opportunity to learn from the failures and successes of different policies to tackle it (even if some countries failed to use these opportunities judiciously).

The implications for our species' ability to resolve a problem like climate change – where the projections are less certain and are likely to hit some regions worse than others; where the costs are paid now but benefits are delayed until far in the future (when we might not even be alive); and where we really need to see a coordinated global effort rather than the piecemeal nationalistic approach that characterised the COVID response – are sobering. But there is still room for cautious optimism.

In fact, our best chance at resolving global public goods problems may be to 'think global, act local'. This catchphrase, coined by the late Nobel laureate Elinor Ostrom, acknowledges the importance of top-down governance and coordination for resolving large-scale complex problems but also highlights the promise of bottom-up or 'polycentric' approaches. As an example, take the Paris Agreement on climate change, formalised in December 2015 and from which President Trump subsequently announced his intention to withdraw in 2017. On the back of this bullish move, more than 3,800 leaders in the US, ranging from governors to CEOs to mayors to university deans, signed a pledge affirming 'We Are Still In', making a commitment to comply with the targets outlined in the Paris Agreement. The latest figures suggest that the signatories to this pledge represent around half of the US total population and economy. Bottom-up approaches build on the foundations that we know are needed for cooperation to succeed. They allow interacting parties to communicate, to develop relationships and to foster trust. They allow for different solutions to be developed for different problems. Crucially, local institutions also allow stakeholders to build consensus and legitimacy around enforcement and penalties for non-compliance.

The importance of thinking global, but acting local is illustrated in the design of quotas for fisheries. A catch-share system incentivises fishers to cooperate, by fishing the stocks sustainably: experts calculate how many fish can be sustainably harvested and each fisher receives an equal share of this quota. Catch-share systems are effective so long as they are perceived to be legitimate: quotas that are distributed unevenly, or are designed without stakeholder input (like the EU Common Fisheries Policy, for example), have a much lower chance of success. To prevent free-riding, compliance often depends on monitoring and sanctioning. Again, legitimacy is key: mutual coercion works, but it has to be mutually agreed upon.

The 2020 coronavirus pandemic offers another illustrative example of thinking global, acting local. To the consternation of many, the UK government initially prevaricated on enacting strict policies or rules mandating social distancing or other means to delay the spread of the disease. For several agonising weeks after the first case appeared in the UK, bars and restaurants remained open, large events and concerts carried on, and people travelled in and out of the country at will, with no measures taken to test people for symptoms or to limit arrivals from countries experiencing high rates of infection. Nevertheless, despite the lack of top-down guidance, many institutions and individuals took early unilateral action: they thought global, but acted local. Major retailers closed their stores before they were forced to; companies mandated that people work from home; several universities (including my own) suspended all face-to-face teaching; and the Premier League cancelled all remaining football matches. These early actions, engineered from the bottom-up, will undoubtedly have saved lives.

*

We may like to think of ourselves as autonomous, self-sufficient creatures, but large-scale crises emphasise that we are fundamentally interdependent. A crisis also offers an opportunity: a moment to pause and take stock of the society we live in, and to consider alternative realities. The Second World War was one of the most profoundly destructive chapters in our species' story, but this terrible event paved the way for some remarkable achievements. Political parties, to a large extent, put aside their disagreements in the interest of a greater good. The 'post-war consensus' was the fertile ground in which the National Health Service was planted, a revolutionary model for society where the state promised to take care of all its citizens, by providing healthcare for all and not just those who could afford it. Wartime also helped to set the wheels of a more gender-equal society into motion. The jobs vacated by men who left to fight were often filled by women: in the US around 6 million women entered the workforce during the war. With a few bumps and starts, this pattern continued after the war ended. Having performed their jobs competently during wartime, there was little substantive argument as to why women should not be allowed to pursue their professional ambitions in peacetime as well.

A pandemic is similar to a war in terms of the societal upheaval it can spark: the ripples from this wave of change can persist far into the future. Some of these ripples can be unwelcome. Authoritarian regimes around the world have used the distraction to further their own interests, introducing knots to the fabric of their democracies that may take years to unravel. But other ripples offer the potential for hope. In the most stringent lockdown periods, air pollution across urban centres plummeted as people's notion of 'essential' journeys rapidly changed. Greenhouse gas emissions temporarily dropped as buzzing flight paths fell silent and coal use in China was estimated to have fallen by around 25% over a four-week period in February and March 2020. At the time of writing, it is too soon to tell whether any of these enforced restrictions might become new habits. Will we continue to work from home, and will we think twice before long-distance travel and stick to virtual conferencing instead?

Some studies suggest this might be possible, and that periods of upheaval are good moments to change our ways. One study where Swiss motorists were offered to exchange their car keys for an e-bike for two weeks reported lasting reductions in car use, even one year after the trial had ended. A similar study involving commuters on the London Underground showed that when a strike forced commuters to alter their route, many of them stuck with the alternative route even after the strike had ended, suggesting that they had been stuck in a suboptimal habit prior to this. A crisis provides the opportunity to rethink our lives and our societies; perhaps offering a silver lining.

One can't help but wonder what the future holds for us and the other occupants of this planet; what life might look like for our own children, and those who will follow. I think we are right to be worried, to declare emergencies and to demand action. But we shouldn't lose hope. Unlike any other species on Earth, we have an ability to find our way out of social dilemmas. We are not simply stuck with the games that nature gives us: we can change the rules. There are countless examples of our ingenuity in this realm. Whether it is hunter-gatherers deciding how to share meat, children deciding to take turns with a toy, or citizens of nation states deciding upon a voting system to elect political representatives into power, creating and changing rules is the means by which we have succeeded at aligning the interests of individuals, allowing them to cooperate to produce greater public goods.

To have a hope of tackling the global problems we face, we need to use these abilities to create effective institutions – rules, agreements and incentives – that favour cooperation and a long-term view over self-interest and short-termism. We can foresee better solutions, we can envision brighter worlds, and we can design the rules of our societies so that people are incentivised to cooperate.

The global human population today stands at nearly 8 billion people, an extraordinary achievement for a species that is no more than an 'acknowledged descendant of an ape'*. For this, we can thank our social instinct – the drive to help our close family, friends and loved ones. Cooperation is undoubtedly the pivotal ingredient in our success. But our enormous presence and impact on the planet now requires us to go beyond instinct and to cooperate in different, less natural ways. It is easy – most of the time – to cooperate with our relatives or within established relationships, but it is far harder to put our faith in people we don't know – and might never meet. Inconveniently, tackling the global problems we now face requires us to do exactly this.

* The eminently quotable Darwin – again.

We undoubtedly have the ability, the technology and the know-how to rise to these challenges. But, tales of societal collapse over history warn us against complacency: failure is a realistic prospect. We should remember that there is no divine plan for our species and no preordained outcome. We still have a chance to get this right, but we won't get another shot at it.

There is an almost fairy-tale quality to the role of cooperation in the human story. If used well it will deliver riches, but in the wrong hands or used in the wrong ways, it will bring ruin. Cooperation has carried us this far in our journey, but if we don't find ways to be better at it – to scale it to the global problems we face – we risk becoming the victims of our own success. Whether this fairy tale has a happy ending is up to us.

ACKNOWLEDGEMENTS

There are a few key people, without whose help this book simply would not have happened. A massive thank you to Sarah-Jayne Blakemore, a friend and mentor, who has helped me in countless ways over the last five years. My editor, Bea Hemming, provided useful feedback and encouragement when I was totally fed up with the book and David Milner worked wonders with copy-editing. Thanks also to my agent, Will Francis, who saw the potential in this book from the start; and to my US editor, Anna De Vries.

A few people read parts or (heroically) the whole of this book and the finished product is immeasurably better for having had their eyes on it. Thanks to Athena Aktipis, Kathleen Ball, Nicole Barbaro, Vaughan Bell, Nigel Bennett, Jonathan Birch, Sinèad English, Helen Haggie, Rebecca Jay, Patrick Kennedy, Dave Lagnado, Elli Leadbeater, Laurent Lehmann and Alicia Melis. A number of people helped me with fact-checking on specific points in the book. Many thanks to Mike Cant, Gemma Clucas, Charlie Cornwallis, Lee Gettler, Nick Lane, Dieter Lukas, Kevin Mitchell, Eleanor Power, Elva Robinson, Jonathan Schulz, Rebecca Sear and Gunter Wagner. Any mistakes are, of course, my own.

A career in science is a bit of a lottery – and failure is the norm. My own career is no exception to this general rule. However, I have been incredibly lucky to have the dice roll in my favour on a few pivotal occasions. In 2003, Mandy Ridley took me on as a research assistant to help her establish the Pied Babbler Research Project.

Given that my only relevant skill at the time was the ability to whistle, I am grateful that she took a punt on me. My supervisor, Tim Clutton-Brock, subsequently encouraged me to apply for a PhD to continue working on the babblers. His judicious use of red pen and terse commentary ('In English next time!') taught me the hard way how to write things people might want to read. In 2008, I attended a conference in the US, where I saw Redouan Bshary give a talk on his cleaner fish work. His first slide was a stunning picture of Lizard Island – like something out of a high-end holiday magazine. I made a point of seeking Redouan out afterwards and mentioning that I wouldn't mind a stint in this tropical paradise. What started as an off-the-cuff remark has blossomed into an enormously fruitful collaboration and friendship, spanning more than a decade. My research has been generously supported by the British Academy, The Leverhulme Trust, the Natural Environment Research Council, the Royal Society, and the Zoological Society of London. In particular, my Royal Society fellowship has given me the academic freedom to pursue my research interests and I will always be grateful for having been a recipient of this award.

One of the things that makes science such a fun job is the people you get to meet and work with. In addition to those mentioned above, I'd like to thank the following mentors, collaborators and friends, many of whom will recognise descriptions of their work in the pages of this book: Quentin Atkinson, Pat Barclay, Louise Barrett, Vaughan Bell, Sandra Binning, Rob Boyd, Lucy Browning, Mike Cant, Innes Cuthill, Nick Davies, Peter Dayan, Joe Devlin, Lee de Wit, Mark Dyble, Jan Engelmann, Gonçalo Faria, Tom Flower, Lucy Foulkes, Simon Gächter, Andy Gardner, Krystyna Golabek, Antonia Hamilton, Uri Hertz, Andy Higginson, Sarah Hodge, Anne Hoggett, Kate Jones, Neil Jordan, Sunjeev Kamboj, Becky Kilner, Sarah Knowles, the late Georgina Mace, Ruth Mace, Marta Manser, Katie McAuliffe, Bonnie Metherell, Kelly Moyes, Mirco Musolesi, Michael Muthukrishna, Martha Nelson-Flower, Ana Pinto, Dominique

Roche, Penny Roth, Andy Russell, Joan Silk, Sarah Smith, Sam Solomon, Anne Sommerfield, Seirian Sumner, Alex Stewart, Rory Sutherland, Alex Thornton, Arne Traulsen, Zegni Triki, Lyle Vail, Gabriella Vigliocco, Stu West, Polly Wiessner, Sharon Wismer and Andy Young. I have also worked with an incredibly talented cohort of students and postdoctoral researchers over the years. While all have helped to shape my thinking, I would especially like to acknowledge Jim Allen, Jack Andrews, Rhea Arini, Joe Barnby, Tommaso Batistoni, Jonathan Bone, Paul Deutchman, Anna Greenburgh, Gabriel Hudson, Oded Keynan, Alice Liefgreen, Alex Thompson, Trisevgeni Papakonstantinou, Elisavet Pappa, Keri Wong and Elena Zwirner.

In the final stages of writing this book my mum passed away, succumbing finally to the illness that had stalked her for the previous seven years. It is hard for me to convey in words what an enormous hole this has left in our family. The last few months have been unimaginably hard and would have been far worse without those key people who offered a safety net: Kim Beazley, Jackie Brown, Hannah Darlington, Sally Grewcock, Jo Harding, Kate Jarman and Bethan Mallen.

I now have personal experience of the costs involved in raising children – and would like to thank my parents and step-parents for their sustained parental (and grandparental) efforts. To my children, who have been a little neglected over the last year or so: I'm sorry and I promise not to write another book.* And to Dave, my endlessly patient husband: a thousand thank yous for always letting me do what I want and never making me feel guilty for it. (And for the morning cups of tea.)

* in 2021.

REFERENCES AND ENDNOTES

CHAPTER EPIGRAPHS

p. 1 'I do not think that there is any evidence that man ever existed as a non-social animal': Charles Darwin, in correspondence to John Morley 14 April 1871, https://www.darwinproject.ac.uk/letter/DCP-LETT-7685.xml.

p. 13 'You can live some sort of life and die without ever hearing the name of Darwin. But if, before you die, you want to understand why you lived in the first place, Darwinism is the one subject that you must study': Richard Dawkins, foreword to John Maynard Smith's *The Theory of Evolution*, Cambridge University Press, 1993.

p. 23 'The individual is, accordingly, a unified commonwealth in which all parts work together for the common end': Rudolf Virchow, *Atoms and Individuals*, 1859.

p. 34 'Mother Nature is a wicked old witch': George C. Williams, *Evolutionary Ethics*, SUNY Press, 1993.

p. 48 'I demur to your saying ... ': Charles Darwin, in correspondence to Neil Arnott, February 1860, https://www.darwinproject.ac.uk/letter/?docId=letters/DCP-LETT-2677.xml.

p. 73 'One touch of nature makes the whole world kin': William Shakespeare, *Troilus and Cressida*, 1602.

p. 81 'In my simplicity, I remember wondering why every gentleman did not become an ornithologist': Charles Darwin, *Autobiographies*, John Murray, 1876–81.

p. 99 'Some pirates achieved immortality by great deeds of cruelty or derring-do ... ': Terry Pratchett, *The Colour of Magic*, Colin Smythe, 1983.

p. 128 'If men were angels, no government would be necessary': James Madison, *The Federalist Papers*, 1788.

p. 141 'Sooner or later everyone sits down to a banquet of consequences': typically attributed to Robert Louis Stevenson from his essay 'Old Mortality', 1884. The original quote is thought to have been longer: 'Books were the proper remedy: books of vivid human import, forcing upon their minds the issues, pleasures, busyness, importance and immediacy of that life in which they stand; books of smiling or heroic temper, to excite or to console; books of a large design, shadowing the complexity of that game of consequences to which we all sit down, the hanger-back not least.'

p. 157 'The sight of a feather in a peacock's tail, whenever I gaze at it, makes me sick!': Charles Darwin, in correspondence to Asa Gray, 1860. https://www.darwinproject.ac.uk/letter/DCP-LETT-2743.xml.

p. 176 'We are evidently evolved to deny that we have evolved to be genetically self-serving': Richard D. Alexander, 'The Challenge of Human Social Behaviour', *Evolutionary Psychology*, Vol. 4, 1–32.

p. 189 'I fully subscribe to the judgement of those writers who maintain that of all the differences between man and the lower animals, the moral sense or conscience is by far the most important': Charles Darwin, *The Descent of Man*, John Murray, 1871.

p. 201 'Man is by nature a political animal': Aristotle, *Politics*, 350 BCE.

p. 208 'Even paranoids have enemies': former Israeli prime minister, Golda Meir (1973), in response to Henry Kissenger, when he accused her of being paranoid for failing to grant additional concessions to the Arabs.

p. 221 'United we stand, divided we fall': Aesop, 'The Four Oxen and the Lion', *Fables*, The Harvard Classics, 1909–14.

p. 235 'Every sin is the result of a collaboration': Stephen Crane, *The Complete Short Stories and Sketches of Stephen Crane*, Doubleday, 2013.

*

INTRODUCTION

p. 3 *The suicidal Brazilian ants* – Tofilski A., Couvillon M. J., Evison S. E. F., Helanterä H., Robinson E. J. H. & Ratnieks F. L. W. 'Preemptive Defensive Self-Sacrifice by Ant Workers'. *The American Naturalist*, 172, E239–43, 2008.

PART 1: THE MAKING OF YOU AND ME

p. 9 *Number of cells in the human body* – Bianconi E., Piovesan A., Facchin F., Beraudi A., Casadei R., Frabetti F., Vitale L., Pelleri, M. C., Tassani, S., Piva, F. & Perez-Amodio, S.

An estimation of the number of cells in the human body'. *Annals of Human Biology,* 40, 463–71, 2013.

p. 9 *the vehicle of reproduction* – This wonderful metaphor was introduced in Dawkins R. *The Selfish Gene*. Oxford University Press, 1976.

p. 10 *Life on Earth timeline* – Bourke A. *Principles of Social Evolution*. Oxford University Press, 2011.

p. 10 *Earth's history scaled down to a single calendar year* – This website is a wonderful resource. https://biomimicry.net/earths-calendar-year/.

p. 11 *Major evolutionary transitions* – Szathmáry E. & Smith J. M. 'The Major Evolutionary Transitions'. *Nature* 374, 227–32, 1995.

p. 11 *Argentine ant super-colony* – Van Wilgenburg E., Torres C. W. & Tsutsui N. D. 'The Global Expansion of a Single Ant Supercolony'. *Evolutionary Applications* 3, 136–43, 2010.

1. A COLD SHUDDER

p. 14 *we can see somewhere in the region of 100,000 to 10 million different colours* – Pointer M. R. & Attridge G. G. 'The Number of Discernible Colours'. *Color Research & Application* 23, 52–4, 1998.

p. 14 *'The eye to this day gives me a cold shudder'* – Charles Darwin in correspondence to Asa Gray, 8 or 9 February 1860. https://www.darwinproject.ac.uk/letter/DCP-LETT-2701.xml.

p. 14 *Darwin's cold shudders* – Darwin's cold shudders may have stemmed not just from the irreducible complexity problem but from the thorny question of how beneficial traits that appeared in the parental generation were passed on to offspring. What were the packages of hereditary information and how were they transmitted? Darwin's grand theory could explain the appearance of design without a designer, but there remained a large missing piece in the puzzle: the mechanism of inheritance, which natural selection relied upon, was elusive.

The most prominent view at the time was that of Jean-Baptiste Lamarck, a French naturalist. Lamarck believed that parents acquired improvements during their lifetime and passed these to offspring; the archetypal example being that of giraffes straining to reach the leaves on higher branches of the tree and passing these elongated necks onto their babies. Though he was aware of Lamarck's ideas, Darwin was unconvinced by them, dismissing Lamarck's book as 'wretched, and one from which I gained nothing'. Instead, Darwin advocated for a kind of 'blending' inheritance, where maternal and paternal features were mixed to give rise to offspring who had intermediate versions of their parents' characters. This also turned out to be problematic because, under this logic, novel beneficial changes are averaged out before selection can act on them. Blending inheritance cannot produce adaptations; instead, it simply washes them away. Eventually Darwin too gave up on this idea, though he was left with nothing to fill the void. He would never understand the mechanics of inheritance.

Unbeknown to Darwin, however, the ways in which character traits are passed onto offspring *had* been discovered, by an Austrian monk named Gregor Mendel. By means of a series of ingenious experiments, using pea plants, Mendel had deduced that hereditary information was stored in particulate form passed from each parent to the offspring in discrete sets of instructions – in what would later become known as genes. Through his breeding experiments, Mendel found he could predict features of the offspring appearance, for example the leaf shape and colour, based on the characters in the parental and grandparental generations. He inferred that these observable traits must be predicted by the packages of information that were passed, unchanged, from one generation to the next. Unfortunately, Darwin never learned of Mendel's experiments (perhaps because they were published in German, in a rather obscure journal) and did not live to witness the ways in which Mendel's insights would be combined with his theory to form the Modern Synthesis: the unified account of how evolution works.

p. 15 *at each step, proved advantageous to the bearer* – This should not be taken to imply that evolution is a directional force, nor that more complex eyes are 'better' than simpler ones. Many living species have these simple photoreceptive cells; for example, single-celled algae use these to detect light and swim towards it.

p. 15 *Genes as packages of information* – Some genes can act like on/off switches, with single variants producing discrete changes in the bearer's appearance and behaviour. For example, blood type is controlled by variation in a single gene and types are discontinuous, in that there are no intermediate forms. Nevertheless, this is the exception rather than the rule. Most traits, from height to shoe size to personality, are under the control of many genes. Gene variants that affect these continuous (or polygenic) traits can therefore be thought of as turning a dial that determines the 'loudness' of that particular characteristic, each variant turning the dial up or down a little, acting in synchrony with all the others. This should not be taken to imply that genes are the *only* factor that influences an individual's phenotype: gene expression can (and often does) depend on environmental factors, meaning that genes can be turned on and off – or up and down – depending on the environment they find themselves in. Moreover, the environment itself can exert large effects on how individuals turn out, sometimes swamping the effects of the underlying genes. For example, a person born to two tall parents may inherit 'tall' gene variants, but if they experience food shortage during development they might not grow to be particularly tall as an adult. Traits like intelligence and proneness to obesity also have a genetic basis, but we can't predict whether individuals will turn out to be overweight geniuses without also considering the environment in which they develop.

p. 16 *As Darwin emphasised, these differences could be slight* – Darwin, C. *On the Origin of Species by Means of Natural Selection*. John Murray, 1859.

p. 16 *Richard Dawkins and 'selfish' genes* – Dawkins, R. *The Selfish Gene*. Oxford University Press, 1976.

p. 18 *Inclusive fitness theory* – Apparently, Stephen Hawking was warned that each equation would halve the readership of his book. Mindful of this fact, I removed equations from the main text. For those who would like to know more about the magically simple equation that underpins inclusive fitness theory, I include it here, with a brief explainer. The key references are Hamilton W. D., 'The Genetical Evolution of Social Behaviour. I', *Journal of Theoretical Biology* 7, 1–16, 1964; Hamilton W. D., 'The Genetical Evolution of Social Behaviour. II', *Journal of Theoretical Biology* 7, 17–52, 1964.

Hamilton's rule states that costly helping behaviour will be favoured by selection whenever $rB - C > 0$. The currency of this equation is genes – and we can perform a simple thought experiment to grasp the logic. Ground squirrels live in groups and often warn one another of danger, by letting out alarm calls if they spot a predator. For simplicity, let's assume that the tendency to scan for predators and give alarm calls is governed by a 'squeaker' gene. This gene results in the bearer performing an action that is costly to itself (standing on its hind legs to scan for predators), but which provides benefits to other squirrels in the vicinity.

To understand how squeaker might persist and spread within a population, we have to consider how the actions of the gene in one squirrel might benefit copies of itself in other bodies. Hamilton's rule tells us that the altruistic actions of squeaker can be favoured by natural selection whenever $rB - C > 0$. The B denotes the benefits to the recipient(s): think of this as how many extra babies another squirrel in the group has because of the actions of the altruistic alarm-caller. C is the cost to the alarm caller herself: the number of babies (and thus descendent copies of squeaker) that she sacrifices in performing the altruistic act. Finally, r is the relatedness: the probability that squeaker is present in the helper's body is also present in the body of the beneficiary, relative to the background probability of sharing that gene at that locus with the average member of the population. Following the logic of this rule, we can predict that altruistic behaviours, like alarm-calling, will be most common in groups comprising primarily kin: a prediction that has been borne out in studies of many species, including ground squirrels.

We can stick some numbers in the equation to better understand how it works. As John Haldane, the evolutionary biologist, famously joked, 'I would gladly give up my life for two brothers, or eight cousins.' There is an evolutionary logic behind this cavalier approach to life. Assuming you have the same parents, you share on average half of your genes with your brother, so $r = 0.5$. You're going to lay down your life for him, so plug in a cost (C) of 2: one for each baby you won't be able to raise (because you're dead). Fortunately, you're saving two brothers, who will have (on average) two babies each, so the benefit (B) = 4. Plugging these numbers into the equation gives us the following: $r(0.5) \times B(4) - C(2) = 0$. Being nitpicky, I'd argue that in this case Haldane should be indifferent as to whether he gives up his life for his two brothers (though I concede that for three full siblings it would be a no-brainer).

p. 19 *Long-tailed tits prefer to help their relatives if their own nest fails* – Leedale A., Sharp S., Simeoni M., Robinson E. & Hatchwell B. 'Fine-scale genetic structure and helping decisions in a cooperatively breeding bird'. *Molecular Ecology* 27, 1714–26, 2018.

p. 20 *Megaponera ants decide who is worth rescuing* – Frank E. T., Wehrhahn M. K. & Linsenmair E. 'Wound Treatment and Selective Help in a Termite-Hunting Ant'. *Proceedings of the Royal Society B: Biological Sciences* 285, 20172457, 2017. Miler K. 'Moribund Ants Do Not Call for Help'. *PLOS ONE* 11, e0151925, 2016.

p. 20 *Childless human couples tended to take in nieces and nephews rather than ageing parents* – Pollet T. V. & Dunbar R. I. M. 'Childlessness predicts helping of nieces and nephews in United States'. *Journal of Biosocial Science* 40, 761–70, 2008.

2. INVENTING THE INDIVIDUAL

p. 23 *Evolutionary perspectives on individuals* – Queller D. C. & Strassmann J. E. 'Beyond Society: The Evolution of Organismality'. *Philosophical Transactions of the Royal Society B: Biological Sciences* 364, no. 1533, 3143–55, 2009. Gardner A. & Grafen A. 'Capturing the Superorganism: A Formal Theory of Group Adaptation'. *Journal of Evolutionary Biology* 22, 659–71, 2009.

p. 24 *to illustrate, let's use a real-life engineering example: the car* – An example chosen to be in keeping with the terminology of replicators and vehicles introduced by Dawkins. Dawkins, R. *The Selfish Gene*. Oxford University Press, 1976.

p. 26 *Extreme morphological specialisation in turtle ants* – Powell S. 'Ecological Specialization and the Evolution of a Specialized Caste in *Cephalotes* Ants'. *Functional Ecology* 22, 902–11, 2008.

p. 26 *Social distancing in ants* – Heinze J. & Bartosz W. 'Moribund Ants Leave Their Nests to Die in Social Isolation'. *Current Biology* 20, 249–52, 2010.

p. 26 *Ants kill infected brood* – Pull C. D., Line V. U., Wiesenhofer F., Grasse A. V., Tragust S., Schmitt T., Brown M. J. F. & Cremer S. 'Destructive Disinfection of Infected Brood Prevents Systemic Disease Spread in Ant Colonies'. *ELife* 7, e32073, 2018.

p. 27 *the colony sweats!* – Ostwald M. M., Smith M. L. & Seeley T. D. 'The Behavioral Regulation of Thirst, Water Collection and Water Storage in Honeybee Colonies'. *Journal of Experimental Biology* 219, 2156–65, 2016.

p. 30 *Individuals can sometimes personally benefit from their involvement in war* – Wiessner P. 'Collective Action for War and for Peace: A Case Study among the Enga of Papua New Guinea'. *Current Anthropology* 60, 224–44, 2019.

p. 30 *The average human has as many microbial cells in their gut as there are cells in the entire body* – Sender R., Fuchs S. & Milo R. 'Revised Estimates for the Number of Human and Bacteria Cells in the Body'. *PLOS Biology* 14, e1002533, 2016.

p. 31 *Stink bug faecal pellets* – Fukatsu T. & Hosokawa T. 'Capsule-Transmitted Gut Symbiotic Bacterium of the Japanese Common Plataspid Stinkbug, *Megacopta Punctatissima*'. *Applied and Environmental Microbiology* 68, 389–96, 2002.

p. 31 *Differences in gut microflora between children delivered vaginally or by Caesarean section* – Salminen S. G., Gibson R., McCartney A. L. & Isolauri E. 'Influence of Mode of Delivery on Gut Microbiota Composition in Seven-Year-Old Children'. *Gut* 53, 1388–9, 2004.

p. 32 *Mitochondria as internal batteries* – Lane N. *Life Ascending: The Ten Great Inventions of Evolution*. Profile Books, 2010.

3. THE RENEGADES WITHIN

p. 35 *Mother's Curse* – Laberge A. M., Jomphe M., Houde L., Vézina H., Tremblay M., Desjardins B., Labuda D., St-Hilaire M., Macmillan C., Shoubridge E. A. & Brais, B. 'A "Fille Du Roy" Introduced the T14484C Leber Hereditary Optic Neuropathy Mutation in French Canadians'. *The American Journal of Human Genetics* 77, 313–7, 2005.

p. 35 *Intragenomic conflict* – Trivers R. & Burt A. *Genes in Conflict: The Biology of Selfish Genetic Elements*. Harvard University Press, 2009.

p. 36 *How intragenomic conflict affects fertility* – Zanders S. E. & Unckless R. L. 'Fertility Costs of Meiotic Drivers'. *Current Biology* 29, R512–20, 2019.

p. 36 *one in seven couples experiences difficulty in conceiving* – https://www.nhs.uk/conditions/infertility/.

p. 37 *The 'parliament of genes'* – Leigh E. G. *Adaptation and Diversity*. Freeman, 1971.

p. 38 *Cancer cases in the US* – Statistics for every year available at https://cancer statisticscenter.cancer.org/.

p. 38 *Hallmark features of cancerous tumours* – Leong S. P., Aktipis A. & Maley C. 'Cancer Initiation and Progression within the Cancer Microenvironment'. *Clinical & Experimental Metastasis* 35, 361–7, 2018.

p. 39 *Tumours as diverse, mutualistic communities* – Tabassum D. P. & Polyak K. 'Tumorigenesis: It Takes a Village'. *Nature Reviews Cancer* 15, 473–83, 2015.

p. 39 *Evolutionary insights to understand and treat cancer* – Aktipis A. *The Cheating Cell*. Princeton University Press, 2020.

PART 2: THE FAMILY WAY

p. 45 *The pain of social exclusion* – Eisenberger N. I., Lieberman M. D. & Williams K. D. 'Does Rejection Hurt? An FMRI Study of Social Exclusion'. *Science* 302, 290–2, 2003.

p. 45 *The negative side effects of loneliness* – Holt-Lunstad J., Smith T. B., Baker M., Harris T. & Stephenson D. 'Loneliness and Social Isolation as Risk Factors for Mortality: A Meta-Analytic Review'. *Perspectives on Psychological Science* 10, 227–37, 2015.

p. 45 *Temperatures inside Emperor penguin huddles* – Gilbert C., Robertson G., Le Maho Y., Naito Y. & Ancel A. 'Huddling Behavior in Emperor Penguins: Dynamics of Huddling'. *Physiology & Behavior* 88, 479–88, 2006.

p. 46 *Evolving multicellularity in a test tube* – Boraas M. E., Seale D. B. & Boxhorn J. E. 'Phagotrophy by a Flagellate Selects for Colonial Prey: A Possible Origin of Multicellularity'. *Evolutionary Ecology* 12, 153–64, 1998.

4. OF MUMS (AND DADS)

p. 49 *The ultimate maternal sacrifice in black-lace spiders* – Kim K.-W. & Horel A. 'Matriphagy in the Spider *Amaurobius Ferox* (*Araneidae, Amaurobiidae*): An Example of Mother-Offspring Interactions'. *Ethology* 104, 1021–37, 1998.

p. 50 *The male-female bond in yellow-billed hornbills* – Finnie M. J. 'Conflict & Communication: Consequences Of Female Nest Confinement In Yellow Billed Hornbills'. PhD thesis, University of Cambridge, 2012.

p. 51 *because a female can spawn her eggs on the male's territory and swim off, leaving him to take care of the kids* – This is known as the 'cruel bind hypothesis', first proposed by Dawkins & Carlisle in 1976. The hypothesis is somewhat contentious, though there are a few studies which support the basic assumptions. Dawkins R. & Carlisle T. R.

'Parental Investment, Mate Desertion and a Fallacy'. *Nature* 262, 131–3, 1976. Czyż B. 'Do Female Penduline Tits *Remiz Pendulinus* Adjust Parental Decisions to Their Mates' Behaviour?'. *Ardea* 99, 27–32, 2011. Kahn A. T., Schwanz L. E. & Kokko H. 'Paternity Protection Can Provide a Kick-Start for the Evolution of Male-Only Parental Care'. *Evolution* 67, 2207–17, 2013.

p. 52 *The metabolic demands of pregnancy trigger birth* – Dunsworth H. M., Warrener A. G., Deacon T., Ellison P. T. & Pontzer H. 'Metabolic Hypothesis for Human Altriciality'. *Proceedings of the National Academy of Sciences* 109, 15212–6, 2012.

p. 52 *The metabolic demands of pregnancy are comparable to those experienced by runners during an ultra-marathon* – https://www.sciencemag.org/news/2019/06/study-marathon-runners-reveals-hard-limit-human-endurance.

p. 52 *Females do most of the parental care in mammals* – Lukas D. & Clutton-Brock T. H. 'Life Histories and the Evolution of Cooperative Breeding in Mammals'. *Proceedings of the Royal Society B: Biological Sciences* 279, 4065–70, 2012.

p. 52 *Paternal care in Hadza & Datoga* – Muller M. N., Marlowe F. W., Bugumba R. & Ellison P. T. 'Testosterone and Paternal Care in East African Foragers and Pastoralists'. *Proceedings of the Royal Society B: Biological Sciences* 276, 347–54, 2009.

p. 53 *Testosterone & paternal care* – Kuo P. X., Braungart-Rieker J. M., Lefever J. E. B., Sarma M. S., O'Neill M. & Gettler L. T. 'Fathers' Cortisol and Testosterone in the Days around Infants' Births Predict Later Paternal Involvement'. *Hormones and Behavior* 106, 28–34, 2018. Gettler L. T., McDade T. W., Feranil A. B. & Kuzawa C. W. 'Longitudinal Evidence That Fatherhood Decreases Testosterone in Human Males'. *Proceedings of the National Academy of Sciences* 108, 16194–9, 2011.

p. 54 *Mate guarding in orb-weaving spiders* – Foellmer M. W. & Fairbairn D. J. 'Spontaneous Male Death during Copulation in an Orb-Weaving Spider'. *Proceedings of the Royal Society B: Biological Sciences* 270, S183–5, 2003.

p. 57 *monogamy came first ... and fathers followed* – Lukas D. & Clutton-Brock T. H. 'The Evolution of Social Monogamy in Mammals'. *Science* 341, 526–30, 2013.

p. 57 *Paternal care in burying beetles* – Hopwood P. E., Moore A. G., Tregenza T. & Royle N. J. 'Male Burying Beetles Extend, Not Reduce, Parental Care Duration When Reproductive Competition Is High'. *Journal of Evolutionary Biology* 28, 1394–402, 2015.

p. 58 *Sex roles in Australia* – Grosjean P. A. & Khattar R. 'It's Raining Men! Hallelujah?'. SSRN Scholarly Paper. Rochester, NY: Social Science Research Network, 2015.

p. 58 *Sex roles in Makushi* – Schacht R. & Borgerhoff Mulder M. 'Sex Ratio Effects on Reproductive Strategies in Humans'. *Royal Society Open Science* 2, 140402, 2015.

5. WORKERS AND SHIRKERS

p. 60 *Parental conflict in zebra finches* – Royle N. J., Hartley I. R. & Parker G. A. 'Sexual Conflict Reduces Offspring Fitness in Zebra Finches'. *Nature* 416, 733–6, 2002.

p. 61 *Theoretical predictions about parental conflict* – McNamara J. M., Székely T., Webb J. N. & Houston A. I. 'A dynamic game-theoretic model of parental care'. *Journal of Theoretical Biology* 205, 605–23, 2000.

p. 62 *Chemical anti-aphrodisiac in burying beetles* – Engel K. C., Stökl J., Schweizer R., Vogel H., Ayasse M., Ruther J. & Steiger S. A. 'Hormone-Related Female Anti-Aphrodisiac Signals Temporary Infertility and Causes Sexual Abstinence to Synchronize Parental Care'. *Nature Communications* 7, 11035, 2016.

p. 63 *More than you ever wanted to know about primate testicles* – Harcourt A. H., Harvey P. H., Larson S. G. & Short R. V. 'Testis Weight, Body Weight and Breeding System in Primates'. *Nature* 293, 55–7, 1981.

p. 66 *The war of the genomes between mother and child* – Haig D. 'Maternal–Fetal Conflict, Genomic Imprinting and Mammalian Vulnerabilities to Cancer'. *Philosophical Transactions of the Royal Society B: Biological Sciences* 370, 20140178, 2015. Haig D. 'Genetic Conflicts in Human Pregnancy'. *The Quarterly Review of Biology* 68, 495–532, 1993.

p. 68 *human-chorionic gonadotropin ... is present in up to 30% of human cancers* – Stenman U. H., Alfthan H. & Hotakainen K. 'Human Chorionic Gonadotropin in Cancer'. *Clinical Biochemistry, Special Issue: Recent Advances in Cancer Biomarkers* 37, 549–61, 2004.

p. 69 *Ectopic pregnancies* – Alkatout I., Honemeyer U., Strauss A., Tinelli A., Malvasi A., Jonat W., Mettler L. & Schollmeyer T. 'Clinical Diagnosis and Treatment of Ectopic Pregnancy'. *Obstetrical & Gynecological Survey* 68, 571–81, 2013. Wang Y.-L., Su T.-H. & Chen H.-S. 'Operative Laparoscopy for Unruptured Ectopic Pregnancy in a Caesarean Scar'. *BJOG: An International Journal of Obstetrics & Gynaecology* 113, 1035–8, 2006.

p. 69 *Evolution of mammalian placenta* – Wildman D. E., Chen C., Erez O., Grossman L. I., Goodman M. & Romero R. 'Evolution of the Mammalian Placenta Revealed by Phylogenetic Analysis'. *Proceedings of the National Academy of Sciences* 103, 3203–8, 2006.

p. 69 *Invasive placentas as a risk-factor for cancer* – Afzal J., Maziarz J. D., Hamidzadeh A., Liang C., Erkenbrack E. M., Nam H., Haeger J. D., Pfarrer C., Hoang T., Ott T. & Spencer, T. 'Evolution of Placental Invasion and Cancer Metastasis Are Causally Linked'. *Nature Ecology & Evolution* 3, 1743–53, 2019.

p. 70 *Prader–Willi syndrome and Angelman syndrome* – Crespi B. & Badcock C. 'Psychosis and Autism as Diametrical Disorders of the Social Brain'. *Behavioral and Brain Sciences* 31, 241–61, 2008.

p. 71 *Fetal microchimerism* – Boddy A. M., Fortunato A., Sayres M. W. & Aktipis A. 'Fetal Microchimerism and Maternal Health: A Review and Evolutionary Analysis of Cooperation and Conflict beyond the Womb'. *BioEssays* 37, 1106–18, 2015.

6. WELCOME TO THE FAMILY

p. 74 *Global distribution of cooperatively breeding species* – Jetz W. & Rubenstein D. R. 'Environmental Uncertainty and the Global Biogeography of Cooperative Breeding in Birds'. *Current Biology* 21, 72–8, 2011. Lukas D. & Clutton-Brock T. 'Climate and the Distribution of Cooperative Breeding in Mammals'. *Royal Society Open Science* 4, 160897, 2017.

p. 75 *The environments in which early humans evolved* – Maslin M. A., Brierley C. M., Milner A. M., Shultz S., Trauth M. H. & Wilson K. E. 'East African Climate Pulses and Early Human Evolution'. *Quaternary Science Reviews* 101, 1–17, 2014.

p. 75 *Diets of humans and other great apes* – Muller M. N., Wrangham R. W. & Pilbeam D. R. *Chimpanzees and Human Evolution*. Harvard University Press, 2018.

p. 76 *Gigantic lions of the Pleistocene* – Manthi F. K., Brown F. H., Plavcan M. J. & Werdelin L. 'Gigantic lion, *Panthera leo,* from the Pleistocene of Natodomeri, eastern Africa'. *Journal of Paleontology, 92,* 305–12, 2018.

p. 76 *Exposure to predation risk associated with increased group size in primate species* – Willems E. P. & van Schaik C. 'The Social Organization of *Homo Ergaster*: Inferences from Anti-Predator Responses in Extant Primates'. *Journal of Human Evolution* 109, 11–21, 2017.

p. 77 *The role of humans in the late Quaternary megafaunal extinctions* – Prescott G. W., Williams D. R., Balmford A., Green R. E. & Manica A. 'Quantitative Global Analysis of the Role of Climate and People in Explaining Late Quaternary Megafaunal Extinctions'. *Proceedings of the National Academy of Science*, 109, 4527–31, 2012.

p. 77 *Prevalence of cooperative breeding across different animal groups – Amphibians*: Doody J. S., Burghardt G. M. & Dinets V. 'Breaking the Social–Non-Social Dichotomy: A Role for Reptiles in Vertebrate Social Behavior Research?'. *Ethology* 119, 95–103, 2013. *Reptiles*: Gardner M. G., Pearson S. K., Johnston G. R. & Schwarz M. P. 'Group Living in Squamate Reptiles: A Review of Evidence for Stable Aggregations'. *Biological Reviews* 91, 925–36, 2016. *Insects*: Otis G. W. 'Sociality of Insects'. *Encyclopedia of Entymology*, 4, 3447–52. Springer, 2008. *Spiders*: Lubin Y. & Bilde T. 'The Evolution of Sociality in

Spiders'. *Advances in the Study of Behavior* 37, 83–145, 2007. *Fish*: Taborsky M. 'Reproductive Skew in Cooperative Fish Groups: Virtue and Limitations of Alternative Modeling Approaches'. *Reproductive Skew in Vertebrates: Proximate and Ultimate Causes*, 265–304. Cambridge University Press, 2009. *Mammals*: Lukas D. & Clutton-Brock T. H. 'The Evolution of Social Monogamy in Mammals'. *Science* 341, 526–30, 2013. *Birds*: Cockburn A. 'Prevalence of Different Modes of Parental Care in Birds'. *Proceedings of the Royal Society B: Biological Sciences* 273, 1375–83, 2006.

p. 78 *When do helpers evolve?* – Lukas D. & Clutton-Brock T. H. 'Climate and the Distribution of Cooperative Breeding in Mammals'. *Royal Society Open Science* 4, 160897, 2017.

p. 78 *mothers have been embedded in vast social networks and children have been raised by multiple caregivers* – Kramer K. L. & Veile A. 'Infant Allocare in Traditional Societies'. *Physiology & Behavior* 193, 117–26, 2018.

p. 79 *This perspective stands in stark contrast to the Western ideal of the nuclear family* – It is worth pointing out that this is more of an ideal than a reality for many. Many families receive *some* help from grandparents (and grandmothers in particular) although the extent of help received varies both within and between countries. Despite the fact that grandparents may be involved, to varying degrees, as helpers in Western families, it is clear that their investments are (for the most part) not as large as they would have been in ancestral human societies. Glaser K., Price D., Montserrat E. R., di Gessa G. & Tinker A. *Grandparenting in Europe: Family Policy and Grandparents' Role in Providing Childcare*, 2018.

p. 79 *Attachment theory* – Fraley R. C. 'Attachment in Adulthood: Recent Developments, Emerging Debates, and Future Directions'. *Annual Review of Psychology* 70, 1375–83, 2019.

p. 80 *National Institute of Child Health and Development study* – National Institute of Child Health and Development Report: https://www.nichd.nih.gov/sites/default/files/publications/pubs/documents/seccyd_06.pdf.

p. 80 *French children … who attended a good nursery typically had fewer emotional and conduct problems* – Gomajee R., El-Khoury F., Côté S., van der Waerden J., Pryor L. & Melchior M. 'Early Childcare Type Predicts Children's Emotional and Behavioural Trajectories into Middle Childhood. Data from the EDEN Mother–Child Cohort Study'. *Journal of Epidemiology and Community Health* 72, 1033–43, 2018.

7. YEARS OF BABBLING

p. 82 If you want to know more about the Kuruman River Reserve mentioned in this chapter (and the research done there), this is a good place to start – http://kalahari-meerkats.com/kmp/.

p. 87 *Sentinel duty in babblers* – Ridley A. R. & Raihani N. J. 'Facultative Response to a Kleptoparasite by the Cooperatively Breeding Pied Babbler'. *Behavioral Ecology* 18, 324–30, 2007.

p. 88 *The watchman's song given by babbler sentinels* – Hollén L. I., Bell M. B. V. & Radford A. N. 'Cooperative Sentinel Calling? Foragers Gain Increased Biomass Intake'. *Current Biology* 18, 576–9, 2008.

p. 89 *Small babbler groups fledge their nestlings earlier* – Raihani N. J., & Ridley A. R. 'Variable Fledging Age According to Group Size: Trade-Offs in a Cooperatively Breeding Bird'. *Biology Letters* 3, 624–7, 2007.

p. 89 *Babblers visit the nest together rather than alone, to protect the nestlings* – Raihani N. J., Nelson-Flower M. J., Moyes K., Browning L. E. & Ridley A. R. 'Synchronous Provisioning Increases Brood Survival in Cooperatively Breeding Pied Babblers'. *Journal of Animal Ecology* 79, 44–52, 2010.

p. 90 *Babbler chicks blackmail adults into feeding them* – Thompson A. M., Raihani N. J., Hockey P. A. R., Britton A., Finch F. M. & Ridley A. R. 'The Influence of Fledgling Location on Adult Provisioning: A Test of the Blackmail Hypothesis'. *Proceedings of the Royal Society B: Biological Sciences* 280, 20130558, 2013.

p. 92 *Young chimps are prolific social learners* – Moore R. 'Social Learning and Teaching in Chimpanzees'. *Biology & Philosophy* 28, 879–901, 2013.

p. 92 *Chimpanzee diet* – Muller M. N., Wrangham R. & Pilbeam D. R. *Chimpanzees and Human Evolution*. 1st edition, Harvard University Press, 2017.

p. 93 *Teaching in Temnothorax ants* – Franks N. R. & Richardson T. 'Teaching in Tandem-Running Ants'. *Nature* 439, 153, 2006.

p. 93 *Teaching in other non-human species* – Thornton A. & Raihani N. J. 'The Evolution of Teaching'. *Animal Behaviour* 75, 1823–36, 2008.

p. 94 *Teaching in meerkats* – Thornton A. & McAuliffe K. 'Teaching in Wild Meerkats'. *Science* 313, 227–9, 2006.

p. 95 *Teaching in babblers* – Raihani N. J. & Ridley A. R. 'Experimental Evidence for Teaching in Wild Pied Babblers'. *Animal Behaviour* 75, 3–11, 2008.

p. 97 *'White lies' in babblers* – Raihani N. J., & Ridley A. R. 'Adult Vocalizations during Provisioning: Offspring Response and Postfledging Benefits in Wild Pied Babblers'. *Animal Behaviour* 74, 1303–9, 2007.

8. IMMORTALS

p. 100 *Division of labour* – Cooper G. A. & West S. A. 'Division of Labour and the Evolution of Extreme Specialization'. *Nature Ecology & Evolution* 2, 1161, 2018. Boomsma

J. J. & Gawne R. 'Superorganismality and Caste Differentiation as Points of No Return: How the Major Evolutionary Transitions Were Lost in Translation'. *Biological Reviews* 93, 28–54, 2018.

p. 101 *we only reliably see extreme and irreversible commitments of this sort in the most supremely social societies, because these societies operate as super-organisms rather than colonies of individuals* – In fact, the evolution of distinct cell types only evolved in three of the twenty-five or so lineages that underwent a transition to multicellularity. These transitions are most likely to occur in large societies (of cells/social insects) where increases in group size can favour increasing specialisation, which in turn permit societies to become even larger and thereby permit even greater specialisation. This kind of positive feedback loop is fragile and can be likened to the engine of an old car: difficult to start and easy to stall. For more, please see: Birch J. 'The Multicellular Organism as a Social Phenomenon'. *The Philosophy of Social Evolution*. Oxford University Press, 2017.

p. 102 *How menopause evolves and why it is unusual* – Ellis S., Franks D. W., Nattrass S., Cant M. A., Bradley D. L., Giles D., Balcomb K. C. & Croft D. P. 'Postreproductive Lifespans Are Rare in Mammals'. *Ecology and Evolution* 8, 2482–94, 2018. Croft D. P., Brent L. J. N., Franks D. W. & Cant M. A. 'The Evolution of Prolonged Life after Reproduction'. *Trends in Ecology & Evolution* 30, 407–16, 2015. Cant M. A. & Johnstone R. A. 'Reproductive Conflict and the Separation of Reproductive Generations in Humans'. *Proceedings of the National Academy of Sciences* 105, 5332–6, 2008.

p. 103 *More sex, later menopause?* – Arnot M. & Mace R. 'Sexual Frequency Is Associated with Age of Natural Menopause: Results from the Study of Women's Health Across the Nation'. *Royal Society Open Science* 7, 191020, 2020.

p. 103 *Human females are born with around 2 million follicles* in situ – Laisk T., Tšuiko O., Jatsenko T., Hõrak P., Otala M., Lahdenperä M., Lummaa V., Tuuri T., Salumets A. & Tapanainen J. S. 'Demographic and Evolutionary Trends in Ovarian Function and Aging'. *Human Reproduction Update* 25, 1–17, 2018.

p. 104 *there are good reasons to believe that dispersal in ancestral humans was female-biased* – These reasons include dispersal patterns in other ape species (chimps, gorillas and bonobos all have female-biased dispersal), patterns we can infer from genetic data, and also what we see in many modern human foraging societies. Summarised in: Cant M. A. & Johnstone R. A. 'Reproductive Conflict and the Separation of Reproductive Generations in Humans'. *Proceedings of the National Academy of Sciences* 105, 5332–6, 2008.

p. 104 *when a grandmother bred alongside her daughter-in-law, all of the children suffered* – Lahdenperä M., Gillespie D. O. S., Lummaa V. & Russell A. F. 'Severe Intergenerational Reproductive Conflict and the Evolution of Menopause'. *Ecology Letters* 15, 1283–90, 2012.

p. 106 *The importance of maternal grandmothers* – Sear R. & Mace R. 'Who Keeps Children Alive? A Review of the Effects of Kin on Child Survival'. *Evolution and Human Behavior* 29, 1–18, 2008. Engelhardt S. C., Bergeron P., Gagnon A., Dillon L. & Pelletier F. 'Using Geographic Distance as a Potential Proxy for Help in the Assessment of the Grandmother Hypothesis'. *Current Biology* 29, 651–6, 2019.

p. 107 *Why grandmothers are (eventually) selected to die* – Chapman S. N., Pettay J. E., Lummaa V. & Lahdenperä M. 'Limits to Fitness Benefits of Prolonged Post-Reproductive Lifespan in Women'. *Current Biology* 29, 1–6, 2019.

p. 108 *Male reproductive opportunities persist into late life* – Vinicius L. & Migliano A. B. 'Reproductive Market Values Explain Post-Reproductive Lifespans in Men'. *Trends in Ecology & Evolution* 31, 172–5, 2016.

p. 109 *Mole-rats* – Bennett N. C. & Faulkes C. G. *African Mole-Rats: Ecology and Eusociality*. Cambridge University Press, 2000.

p. 110 *Secondary growth in mole-rat queens* – Young A. J. & Bennett N. C. 'Morphological Divergence of Breeders and Helpers in Wild Damaraland Mole-Rat Societies'. *Evolution* 64, 3190–7, 2010.

p. 111 *Mole-rat queens have longer lifespans* – Graham R. J., Smith M. & Buffenstein R. 'Naked Mole-Rat Mortality Rates Defy Gompertzian Laws by Not Increasing with Age'. *ELife* 7, e31157, 2018.

p. 111 *Reproduction-survival trade-offs* – Blacher P., Huggins T. J. & Bourke A. F. G. 'Evolution of Ageing, Costs of Reproduction and the Fecundity–Longevity Trade-off in Eusocial Insects'. *Proceedings of the Royal Society B: Biological Sciences* 284, 20170380, 2017.

p. 111 *Some ants defy the reproduction-survival trade-offs (they are known as 'Darwinian demons')* – Schrempf A., Giehr J., Röhrl R., Steigleder S. & Heinze J. 'Royal Darwinian Demons: Enforced Changes in Reproductive Efforts Do Not Affect the Life Expectancy of Ant Queens'. *The American Naturalist* 189, 436–42, 2017.

p. 112 *How to evolve a long lifespan* – Healy K., Guillerme T., Sive F., Kane A., Kelly S., McClean D., Kelly D. J., Donohue I., Jackson A. L. & Cooper N. 'Ecology and Mode-of-Life Explain Lifespan Variation in Birds and Mammals'. *Proceedings of the Royal Society B: Biological Sciences* 281, 1784, 20140298, 2014. Keller L. & Genoud M. 'Extraordinary Lifespans in Ants: A Test of Evolutionary Theories of Ageing'. *Nature* 389, 958–60, 1997.

p. 112 *Intensive care units in ants* – Stroeymeyt N., Grasse A. V., Crespi A., Mersch D.P., Cremer S. & Keller L. 'Social Network Plasticity Decreases Disease Transmission in a Eusocial Insect'. *Science* 362, 941–5, 2018.

9. ASCENDING THE THRONE

p. 115 *it is thought that Kim Jong-un first met his half-brother just five years before the latter was murdered* – https://en.wikipedia.org/wiki/Kim_Jong-nam.

p. 116 *Eviction and abortion in meerkats* – Young A. J., Carlson A. A., Monfort S. L., Russell A. F., Bennett N. C. & Clutton-Brock T. 'Stress and the Suppression of Subordinate Reproduction in Cooperatively Breeding Meerkats'. *Proceedings of the National Academy of Sciences* 103, 12005–10, 2006.

p. 116 *Meerkat females as the world's worst grandmothers* – Lukas D. & Huchard E. 'The Evolution of Infanticide by Females in Mammals'. *Philosophical Transactions of the Royal Society B: Biological Sciences* 374, 20180075, 2019.

p. 116 *When scientists put banded mongooses on the contraceptive pill* – Cant M. A., Nichols H. J., Johnstone R. A. & Hodge S. J. 'Policing of Reproduction by Hidden Threats in a Cooperative Mammal'. *Proceedings of the National Academy of Sciences* 111, 326–30, 2014.

p. 117 *Breeding synchrony in banded mongooses* – Hodge S. J., Bell M. B. V. & Cant M. A. 'Reproductive Competition and the Evolution of Extreme Birth Synchrony in a Cooperative Mammal'. *Biology Letters* 7, 54–6, 2011.

p. 117 *Dominance relations in* Polistes *wasps* – West-Eberhard M. J. 'Dominance Relations in *Polistes Canadensis* (L.), a Tropical Social Wasp'. *Monitore Zoologico Italiano-Italian Journal of Zoology* 20, 263–81, 1986.

p. 120 *Wasp workers kill their mother* – Loope K. J. 'Queen Killing Is Linked to High Worker-Worker Relatedness in a Social Wasp'. *Current Biology* 25, 2976–9, 2015.

10. THE SOCIAL DILEMMA

p. 128 *The classic* Golden Balls *episode, featuring Stephen and Sarah* – *Golden Balls* (ITV, 14 March 2008). https://www.youtube.com/watch?v=7FbkwrhW_0I.

p. 129 *Prisoner's Dilemma* – Strictly, for a two-player interaction to be a Prisoner's Dilemma then T > R > P > S. T is the 'temptation to cheat': the pay-off you get from exploiting a cooperative partner. R is the 'reward' for mutual cooperation. P is the 'punishment' for mutual defection, and S is the 'sucker's pay-off': the pay-off you get if you cooperate and the partner defects. The Prisoner's Dilemma used in Split or Steal is actually not strictly a Prisoner's Dilemma since P = S, rather than P > S.

p. 130 *Brain scans reveal the warm glow of giving* – Harbaugh W. T., Mayr U. & Burghart D. R. 'Neural Responses to Taxation and Voluntary Giving Reveal Motives for Charitable Donations'. *Science* 316, 1622–5, 2007.

p. 130 *Toddlers are happier when they share* – Aknin L. B., Hamlin J. K. & Dunn E. W. 'Giving Leads to Happiness in Young Children'. *PLOS ONE* 7, e39211, 2012.

p. 130 *Spending money on others increases happiness* – Dunn E. W., Aknin L. B. & Norton M. I. 'Prosocial Spending and Happiness: Using Money to Benefit Others Pays Off'. *Current Directions in Psychological Science* 23, 41–7, 2014.

p. 130 *and improves cardiovascular health* – Whillans A. V., Dunn E. W., Sandstrom G. M., Dickerson S. S. & Madden K. M. 'Is Spending Money on Others Good for Your Heart?'. *Health Psychology* 35, 574–83, 2016.

p. 131 *The 'identifiable victim effect'* – Jennni K. & Loewenstein G. 'Explaining the Identifiable Victim Effect'. *Journal of Risk and Uncertainty* 14, 235–57, 1997.

p. 000 *Human empathy is bounded and fickle: it is only bestowed upon 'worthy' individuals* – One study, done in India, found that people were less likely to help an identifiable victim if the victim was perceived to be low-caste. Work I have done with Laura Thomas-Walters also found that identifiable victims probably have to be humans. We ran a study where we presented a conservation appeal (asking people to donate to a conservation charity) with a picture of a single 'identifiable victim', for example a single polar bear, or with a picture of a group of animals. The pictures of an identifiable victim were not more effective at soliciting donations than the group pictures (people cared more about the animals we used, donating more when we showed them 'cute' endangered species, rather than the uglier, but still endangered, creatures). Deshpande A. & Spears D. 'Who Is the Identifiable Victim? Caste and Charitable Giving in Modern India'. *Economic Development and Cultural Change* 64, 299–321, 2016. Thomas-Walters L. & Raihani N. J. 'Supporting Conservation: The Roles of Flagship Species and Identifiable Victims'. *Conservation Letters* 10, 581–7, 2017.

p. 132 *Proximate and ultimate explanations* – Tinbergen N. 'On Aims and Methods of Ethology'. *Ethology* 20, 410–33, 1963.

p. 132 *Altruists' brains are different* – Marsh A. A., Stoycos S. A., Brethel-Haurwitz K. M., Robinson P., VanMeter J. W. & Cardinale E. M. 'Neural and Cognitive Characteristics of Extraordinary Altruists'. *Proceedings of the National Academy of Sciences* 111, 15036–41, 2014.

p. 134 *Reciprocity* – Trivers R. L. 'The Evolution of Reciprocal Altruism'. *The Quarterly Review of Biology* 46, 35–57, 1971.

p. 134 *The other species that solve cooperation problems with reciprocity* – Raihani N. J. & Bshary R. 'Resolving the Iterated Prisoner's Dilemma: Theory and Reality. *Journal of Evolutionary Biology* 24, 1628–39, 2011.

p. 135 *Hamlet fish* – Fischer E. A. 'The Relationship between Mating System and Simultaneous Hermaphroditism in the Coral Reef Fish, *Hypoplectrus Nigricans* (Serranidae)'. *Animal Behaviour* 28, 620–33, 1980.

p. 138 *Interdependence* – Roberts G. 'Cooperation through Interdependence'. *Animal Behaviour* 70, 901–8, 2005. Aktipis A., Cronk L., Alcock J., Ayers J. D., Baciu C., Balliet D., Boddy A. M., Curry O. S., Krems J. A., Muñoz, A. & Sullivan, D. 'Understanding Cooperation through Fitness Interdependence'. *Nature Human Behaviour* 2, 429–31, 2018.

11. AN EYE FOR AN EYE

p. 142 *Punishment in the Turkana* – Mathew S. & Boyd R. 'Punishment Sustains Large-Scale Cooperation in Prestate Warfare'. *Proceedings of the National Academy of Sciences* 108, 11375–80, 2011. Mathew S. & Boyd R. 'The Cost of Cowardice: Punitive Sentiments towards Free Riders in Turkana Raids'. *Evolution and Human Behavior* 35, 58–64, 2014.

p. 143 *The staff-room kitchen sink experiment* – Raihani N. J. & Hart T. 'Free-Riders Promote Free-Riding in a Real-World Setting'. *Oikos*, 119, 1391–3, 2010.

p. 145 *Laboratory studies of punishment* – Fehr E. & Gächter S. 'Altruistic Punishment in Humans'. *Nature* 415, 137–40, 2002.

p. 145 *Punishment doesn't always convert defectors into cooperators* – Raihani N. J & Bshary R. 'Punishment: One Tool, Many Uses'. *Evolutionary Human Sciences* 1, e12, 2019.

p. 147 *Punishment is a service that is costly to provide* – Raihani N. J., Thornton A. & Bshary R. 'Punishment and Cooperation in Nature'. *Trends in Ecology & Evolution* 27, 288–95, 2012.

p. 148 *Punishing others feels good* – de Quervain D. J. F., Fischbacher U., Treyer V., Schellhammer M., Schnyder U., Buck A. & Fehr E. 'The Neural Basis of Altruistic Punishment'. *Science* 305, 1254–58, 2004.

p. 148 *Children enjoyed watching puppet punishment* – Mendes N., Steinbeis N., Bueno-Guerra N., Call J. & Singer T. 'Preschool Children and Chimpanzees Incur Costs to Watch Punishment of Antisocial Others'. *Nature Human Behaviour* 2, 45–51, 2018.

p. 150 *Cleaner fish prefer to eat client mucus, rather than ectoparasites* – Grutter A. S. & Bshary R. 'Cleaner Wrasse Prefer Client Mucus: Support for Partner Control Mechanisms in Cleaning Interactions'. *Proceedings of the Royal Society B: Biological Sciences* 270, S242–4, 2003.

p. 154 *Male cleaner fish punish cheating females* – Raihani N. J., Grutter A. S. & Bshary R. 'Punishers Benefit From Third-Party Punishment in Fish'. *Science* 327, 171, 2010.

p. 155 *Punishers can acquire a reputation for being formidable or trustworthy* – Raihani N. J. & Bshary R. 'The Reputation of Punishers'. *Trends in Ecology & Evolution* 30, 98–103, 2015. Barclay P. 'Reputational Benefits for Altruistic Punishment'. *Evolution and Human Behavior* 27, 325–44, 2006. Raihani N. J. & Bshary R. 'Third-Party Punishers Are Rewarded, but Third-Party Helpers Even More So'. *Evolution* 69, 993–1003, 2015.

12. PEACOCKING

p. 158 *Incentivising energy-efficient behaviour through signalling* – Yoeli E., Hoffman M., Rand D. G. & Nowak M. A. 'Powering up with Indirect Reciprocity in a Large-Scale Field Experiment'. *Proceedings of the National Academy of Sciences* 110, 10424–9, 2013.

p. 158 *Voter turnout in Switzerland* – Funk P. 'Social Incentives and Voter Turnout: Evidence From the Swiss Mail Ballot System'. *Journal of the European Economic Association* 8, 1077–103, 2010.

p. 159 *Choosy client fish prefer honest cleaner fish* – Bshary R. & Schäffer D. 'Choosy Reef Fish Select Cleaner Fish That Provide High-Quality Service'. *Animal Behaviour* 63, 557–64, 2002.

p. 159 *Cleaner fish give better service when they are being watched* – Pinto A., Oates J., Grutter A. S. & Bshary R. 'Cleaner Wrasses *Labroides Dimidiatus* Are More Cooperative in the Presence of an Audience'. *Current Biology* 21, 1140–4, 2011.

p. 159 *Even humans struggle with reputation management until we hit middle childhood* – Engelmann J. M. & Rapp D. J. 'The Influence of Reputational Concerns on Children's Prosociality'. *Current Opinion in Psychology* 20, 92–5, 2018.

p. 159 *Great apes do not know or care what others think of them* – Tomasello M. *Becoming Human: A Theory of Ontogeny*. Harvard University Press, 2019.

p. 162 *Dark Web sales data* – https://rstudio-pubs-static.s3.amazonaws.com/279562_48fcbe87ec814596944fb8bb59b10ac3.html.

p. 162 *Prison gangs as trading organisations* – Skarbek D. 'Prison Gangs, Norms, and Organizations'. *Journal of Economic Behavior & Organization* 82, 96–109, 2012.

p. 163 *Maghribi traders* – Greif A. 'Reputation and Coalitions in Medieval Trade: Evidence on the Maghribi Traders'. *The Journal of Economic History* 49, 857–82, 1989.

p. 165 *False signalling by cleaner fish* – Bshary R. 'Biting Cleaner Fish Use Altruism to Deceive Image-Scoring Client Reef Fish'. *Proceedings of the Royal Society B: Biological Sciences* 269, 2087–93, 2002.

p. 168 *Meriam turtle hunting* – Smith E. A. & Bliege Bird R. L. 'Turtle Hunting and Tombstone Opening: Public Generosity as Costly Signaling'. *Evolution and Human Behavior* 21, 245–61, 2000.

p. 169 *the best hunters in a group also performed better on a battery of tasks aimed at testing hunting-relevant skills* – Stibbard-Hawkes D. N. E., Attenborough R. D. & Marlowe F. W. 'A Noisy Signal: To What Extent Are Hadza Hunting Reputations Predictive of Actual Hunting Skills?'. *Evolution and Human Behavior* 39, 639–51, 2018.

p. 170 *The social benefits of generosity* – Bliege Bird R. L. & Power E. A. 'Prosocial Signaling and Cooperation among Martu Hunters'. *Evolution and Human Behavior* 36, 389–97, 2015. Gurven M., Allen-Arave W., Hill K. & Hurtado M. '"It's a Wonderful Life": Signaling Generosity among the Ache of Paraguay'. *Evolution and Human Behavior* 21, 263–82, 2000.

p. 171 *People prefer poor-but-fair partners over those who were rich-but-stingy* – Raihani N. J. & Barclay P. 'Exploring the Trade-off between Quality and Fairness in Human Partner Choice'. *Royal Society Open Science* 3, 160510, 2016.

p. 171 *What men and women prioritise in mating partners* – Buss D. M. 'Sex Differences in Human Mate Preferences: Evolutionary Hypotheses Tested in 37 Cultures'. *Behavioral and Brain Sciences* 12, 1–49, 1989. Conroy-Beam D. & Buss D. M. 'Why Is Age so Important in Human Mating? Evolved Age Preferences and Their Influences on Multiple Mating Behaviors'. *Evolutionary Behavioral Sciences* 13, 127–57, 2019.

p. 174 *Men compete when donating on online fundraising pages* – Raihani N. J. & Smith S. 'Competitive Helping in Online Giving'. *Current Biology* 25, 1183–6, 2015.

p. 175 *Anonymous donations on online fundraising pages* – Raihani N. J. 'Hidden Altruism in a Real-World Setting'. *Biology Letters* 10, 20130884, 2014.

13. THE REPUTATION TIGHTROPE

p. 176 *Intuitive bullshit detection in young children* – Heyman G. D., Fu G. & Lee K. 'Evaluating Claims People Make About Themselves: The Development of Skepticism'. *Child Development* 78, 367–75, 2007.

p. 177 *Children under the age of five don't really know or care what others think of them* – Engelmann J. M., Herrmann E. & Tomasello M. 'Five-Year Olds, but Not Chimpanzees, Attempt to Manage Their Reputations'. *PLOS ONE* 7, e48433, 2012.

p. 179 *Do-gooder derogation* – Monin B., Sawyer P. J. & Marquez M. J. 'The Rejection of Moral Rebels: Resenting Those Who Do the Right Thing'. *Journal of Personality and Social Psychology* 95, 76–93, 2008. Parks C. D. & Stone A. B. 'The Desire to Expel Unselfish Members from the Group'. *Journal of Personality and Social Psychology* 99, 303–10, 2010.

p. 179 *Antisocial punishment* – Herrmann B., Thöni C. & Gächter S. 'Antisocial Punishment Across Societies'. *Science* 319, 1362–67, 2008. Raihani N. J. & Bshary R. 'Punishment: One Tool, Many Uses'. *Evolutionary Human Sciences* 1, e12, 2019.

p. 179 *Anonymous donations on online fundraising pages* – Raihani N. J. 'Hidden Altruism in a Real-World Setting'. *Biology Letters* 10, 20130884, 2014.

p. 180 *Tainted altruism* – Newman G. E. & Cain D. M. 'Tainted Altruism'. *Psychological Science* 25, 648–655, 2014.

p. 182 *How advertising good deeds can completely backfire* – Lee R. B. 'Eating Christmas in the Kalahari'. *Natural History*, December 1969.

p. 183 *How costly acts of devotion affect reputation in rural South India* – Power E. A. & Ready E. 'Building Bigness: Reputation, Prominence, and Social Capital in Rural South India'. *American Anthropologist* 120, 444–59, 2018.

p. 184 *Power's work also highlights the importance of investment into subtle social signals* – Bird R. B., Ready E. & Power E. A. 'The Social Significance of Subtle Signals'. *Nature Human Behaviour* 2, 452, 2018.

14. FACEBOOK FOR CHIMPS

p. 189 *Lottery-winners' neighbours were more likely to run up their debts and file for bankruptcy* – Agarwal S., Mikhed V. & Scholnick B. 'Does the Relative Income of Peers Cause Financial Distress? Evidence from Lottery Winners and Neighboring Bankruptcies'. *Federal Reserve Bank of Philadelphia Working Papers*, 2018.

p. 190 *The wealthiest Americans classify themselves as middle class* – Shenker-Osorio A. 'Why Americans All Believe They Are "Middle Class"'. *The Atlantic*, 2013. https://www.theatlantic.com/politics/archive/2013/08/why-americans-all-believe-they-are-middle-class/278240/.

p. 190 *How social comparison can damage self-esteem* – Kross E., Verduyn P., Demiralp E., Park J., Lee D. S., Lin N., Shablack H., Jonides J. & Ybarra O. 'Facebook Use Predicts Declines in Subjective Well-Being in Young Adults'. *PLOS ONE* 8, e69841, 2013.

p. 190 *Unfair offers are typically rejected* – Camerer C. F. *Behavioral Game Theory: Experiments in Strategic Interaction*. Princeton University Press, 2011.

p. 190 *Economics students as rational maximisers* – Carter J. R. & Irons M. D. 'Are Economists Different, and If So, Why?'. *Journal of Economic Perspectives* 5, 171–7, 1991. Cipriani G. P., Lubian D. & Zago A. 'Natural Born Economists?'. *Journal of Economic Psychology* 30, 455–68, 2009.

p. 191 *Cross-cultural differences in what is perceived as being 'fair'* – Henrich J., Boyd R., Bowles S., Camerer C., Fehr E., Gintis H., McElreath R., Alvard M., Barr A., Ensminger J. & Henrich N. S. '"Economic Man" in Cross-Cultural Perspective: Behavioral Experiments in 15 Small-Scale Societies'. *Behavioral and Brain Sciences* 28, 795–815, 2005.

p. 191 *Children reject unequal outcomes, even at a personal cost* – McAuliffe K., Blake P. R. & Warneken F. 'Children Reject Inequity out of Spite'. *Biology Letters* 10, 20140743, 2014.

p. 192 *Do non-human primates have a sense of fairness?* – To test whether non-human species dislike unfair outcomes we need experiments that allow us to get at these preferences, without asking animals how they feel about it and without being able to explain the rules of an arbitrary experiment to them. To do this, in 2003, Sarah Brosnan and Franz de Waal designed an ingenious task, called the inequity aversion game, and the first creature they tested was a capuchin monkey. Prior to taking part in the experiments, the capuchins were trained to trade tokens with a human experimenter in exchange for a food reward, something they can learn to do with time and patience in the lab. During the experiment, two monkeys sat in adjacent cages, in full sight of one another, and exchanged tokens with the experimenter in return for food rewards. At first, they both received cucumber pieces, for which they happily exchanged their tokens. But then, the experimenter began rewarding the monkeys differently for their tokens, continuing to provide one individual with cucumber, but upgrading the partner to a better food item: a grape. Upon realising that they had become disadvantaged relative to the partner, monkeys ostensibly 'refused unequal pay' by ceasing to trade with the experimenter, and even flinging the offending piece of cucumber back in the experimenter's face. Experiments with chimps report similar results: an individual who gets unequal pay will throw the chimpanzee equivalent of a temper tantrum and refuse to trade further with the experimenter.

At first blush, these results are suggestive of evidence for inequity aversion. But, there are a couple of problems with accepting this interpretation carte blanche. One issue is that these results haven't held up terribly well when other researchers have tried to replicate them. Another is that there are some unfortunate features of the experimental design that make it hard to infer what is going on. One ironic aspect of the task is that the capuchin (or chimp) who refuses cucumber when the partner gets grape actually *increases* the unfairness of the situation, rather than reducing it. This is unlike the experiments with children and adults, where refusals also deprive the partner of their rewards. Indeed, further experiments have shown that children will not reject a low-quality item unless they can also deprive the partner of the higher-quality prize. In other words, unlike capuchins, children don't 'throw the cucumber'. In addition, many of the 'positive' results of inequity aversion in non-human primate species are also consistent with a simpler interpretation: subjects have expectations about the reward they will get, and are annoyed when they are presented with something inferior. A fairness preference is social by definition; it means you have to compare your own pay-offs to someone else's. Non-human primates don't seem to perform this social comparison step, but instead evaluate what they get relative to what is theoretically available. Brosnan S. F. & de Waal F. B. M. 'Monkeys Reject Unequal Pay'. *Nature* 425, 297–9, 2003. McAuliffe K., Chang L. W., Leimgruber K. L., Spaulding R., Blake P. R. & Santos L. 'Capuchin Monkeys, *Cebus Apella*, Show No Evidence for Inequity Aversion in a Costly Choice Task'. *Animal Behaviour* 103, 65–74, 2015. Roma P. G., Silberberg A., Ruggiero A. M. & Suomi S. J. 'Capuchin Monkeys, Inequity Aversion, and the Frustration Effect'. *Journal of Comparative Psychology* 120, 67–73, 2006. Silberberg A., Crescimbene L., Addessi E., Anderson J. R. & Visalberghi E. 'Does Inequity Aversion Depend on a Frustration Effect? A Test with Capuchin Monkeys (*Cebus Apella*)'. *Animal Cognition* 12, 505–9, 2009. Bräuer J., Call J. & Tomasello M. 'Are apes inequity averse? New data on the token-exchange paradigm'. *American Journal of Primatology* 71, 175–81, 2009. Bräuer J., Call J. &Tomasello M. 'Are Apes Really Inequity Averse?'. *Proceedings of the Royal Society B: Biological Sciences* 273, 3123–8, 2006. Ulber J., Hamann K. & Tomasello M. 'Young Children, but Not Chimpanzees, Are Averse to Disadvantageous and Advantageous Inequities'. *Journal of Experimental Child Psychology* 155, 48–66, 2017. Kaiser I., Jensen K., Call K. & Tomasello M. 'Theft in an Ultimatum Game: Chimpanzees and Bonobos Are Insensitive to Unfairness'. *Biology Letters* 8, 942–45, 2012. Jensen K., Call J. & Tomasello M. 'Chimpanzees Are Rational Maximizers in an Ultimatum Game'. *Science* 318, 107–9, 2007.

p. 193 *each of our ancestors was, in effect, on a camping trip that lasted a lifetime* – Cosmides L. & Tooby J. *Evolutionary Psychology: A Primer*. Center for Evolutionary Psychology, University of California Santa Barbara, 1997.

p. 193 *The social universe of hunter-gatherers* – Hill K. R., Wood B. M., Baggio J., Hurtado A. M. & Boyd R. 'Hunter-Gatherer Inter-Band Interaction Rates: Implications for Cumulative Culture'. *PLOS ONE* 9, e102806, 2014.

p. 194 *Energy requirements of brains* – Raichle M. E. & Gusnard D. A. 'Appraising the Brain's Energy Budget'. *Proceedings of the National Academy of Sciences* 99, 10237–9, 2002.

p. 194 *The energetics of chess* – Kumar A. 'The Grandmaster Diet: How to Lose Weight While Barely Moving'. *ESPN*, 13 September 2019.

p. 195 *Chimpanzee hunting: teamwork or not?* – It is important to note that all observations of purportedly cooperative hunting derive from a population of chimps in the Taï National Forest, in the Ivory Coast, whereas the contradictory results have been obtained from a population in Tanzania. This raises the possibility that there could be important differences in socio-cognitive and behavioural strategies among Eastern and Western chimp populations but until now this idea remains untested.

p. 195 *Chimps that hunt together increase their own personal chance of success* – Gilby I. C., Machanda Z. P., Mjungu D. C., Rosen J., Muller M. N., Pusey A. E. & Wrangham R. '"Impact Hunters" Catalyse Cooperative Hunting in Two Wild Chimpanzee Communities'. *Philosophical Transactions of the Royal Society B: Biological Sciences* 370, 20150005, 2015.

p. 196 *Children like to work together, chimps prefer to work alone* – Rekers Y., Haun D. B. M. & Tomasello M. 'Children, but Not Chimpanzees, Prefer to Collaborate'. *Current Biology* 21, 1756–8, 2011.

p. 197 *Some evidence that chimps from Taï National Park preferentially share meat with hunters that were instrumental in the hunt's success* – Samuni L., Preis A., Deschner T., Crockford C. & Wittig R. M. 'Reward of Labor Coordination and Hunting Success in Wild Chimpanzees'. *Communications Biology* 1, 138, 2018.

p. 197 *But experiments in captive chimps show that whether you get rewarded or not depends on whether you are close to the food-holder, not whether you helped to obtain the food* – John M., Duguid S., Tomasello M. & Melis A. P. 'How Chimpanzees (*Pan Troglodytes*) Share the Spoils with Collaborators and Bystanders'. *PLOS ONE* 14, e0222795, 2019.

p. 197 *Humans (but not chimps) understand helpful gestures* – Tomasello M. *Becoming Human: A Theory of Ontogeny*. Harvard University Press, 2019.

p. 199 *Spontaneous helping in human children* – Warneken F. & Tomasello M. 'Altruistic Helping in Human Infants and Young Chimpanzees'. *Science* 311, 1301–3, 2006.

p. 199 *Chimps don't reliably help a partner* – Silk J. B., Brosnan S. F., Vonk J., Henrich J., Povinelli D. J., Richardson A. S., Lambeth S. P., Mascaro J. & Schapiro S. J. 'Chimpanzees are indifferent to the welfare of unrelated group members'. *Nature* 437, 1357–1359, 2005.

15. MUTINY

p. 202 *Predatory ship captains and pirate constitutions* – Leeson P. T. 'An-*arrgh*-chy: The Law and Economics of Pirate Organization'. *Journal of Political Economy* 115, 1049–94, 2007.

p. 204 *The importance of friendships* – Dunbar R. I. M. 'The Anatomy of Friendship'. *Trends in Cognitive Sciences* 22, 32–51, 2018.

p. 204 *BaYaka people with more friends have higher BMI* – Chaudhary N., Salali G. D., Thompson, J., Rey A., Gerbault P., Stevenson E. G. J., Dyble M., Page A. E., Smith D., Mace R. & Vinicius L. 'Competition for Cooperation: Variability, Benefits and Heritability of Relational Wealth in Hunter-Gatherers'. *Scientific Reports* 6, 29120, 2016.

p. 205 *Friends increased survival in Andersonville* – Costa D. L. & Kahn M. E. 'Surviving Andersonville: The Benefits of Social Networks in POW Camps'. *NBER Working Paper*, 11825, 2005.

p. 205 *Baboons benefit when they form friendships with unrelated males* – Silk, J. B., Alberts S. C. & Altmann J. 'Social Bonds of Female Baboons Enhance Infant Survival'. *Science* 302, 1231–34, 2003.

p. 205 *Friendships seem to be especially important for chimpanzee males* – Alliances seem to be less important for female chimps because females largely compete over food, rather than access to mates. They don't need to team up to solve this problem: simply avoiding one another can be an effective strategy. Indeed, female chimpanzees are reported to spend around 65% of their time foraging either alone or with their infant. Muller M. N., Wrangham R. W. & Pilbeam D. R. *Chimpanzees and Human Evolution*. Harvard University Press, 2018.

p. 205 *The story of Kasonta, Sobongo and Kamemanfu* – Nishida T. 'Alpha Status and Agonistic Alliance in Wild Chimpanzees (*Pan Troglodytes Schweinfurthii*)'. *Primates* 24, 318–36, 1983.

p. 206 *Chimps intervene in grooming interactions between others* – Mielke A., Samuni L., Preis A., Gogarten J. F., Crockford C. & Wittig R. M. 'Bystanders Intervene to Impede Grooming in Western Chimpanzees and Sooty Mangabeys'. *Royal Society Open Science* 4, 171296, 2017.

p. 207 *We reflexively categorise others as 'in-group' or 'out-group'* – Dunham Y. 'Mere Membership'. *Trends in Cognitive Sciences* 22, 780–93, 2018.

16. HERE BE DRAGONS

p. 208 *Bedlam patient James Tilly Matthews* – Carpenter P. K. 'Descriptions of Schizophrenia in the Psychiatry of Georgian Britain: John Haslam and James Tilly Matthews'. *Comprehensive Psychiatry* 30, 332–8, 1989.

p. 209 *The prevalence of paranoia in the general population* – Freeman D. S., McManus S., Brugha T., Meltzer H., Jenkins R. & Bebbington P. 'Concomitants of Paranoia in the General Population'. *Psychological Medicine* 41, 923–36, 2011.

p. 209 *Paranoia as a feature, rather than just a bug, in our psychology* – Raihani N. J. & Bell V. 'An Evolutionary Perspective on Paranoia'. *Nature Human Behaviour* 3, 114–21, 2019.

p. 211 *The social-safety index* – Boyer P., Firat R. & van Leeuwen F. 'Safety, Threat, and Stress in Inter-group Relations: A Coalitional Index Model'. *Perspectives on Psychological Science* 10, 434–50, 2015.

p. 211 *Studies that suggest paranoia develops in response to social threat* – Raihani N. J. & Bell V. 'An Evolutionary Perspective on Paranoia'. *Nature Human Behaviour* 3, 114–21, 2019. Gayer-Anderson C. & Morgan C. 'Social Networks, Support and Early Psychosis: A Systematic Review'. *Epidemiology and Psychiatric Sciences* 22, 131–46, 2013. Catone G., Marwaha S., Kuipers E. & Lennox B. 'Bullying Victimisation and Risk of Psychotic Phenomena: Analyses of British National Survey Data'. *The Lancet Psychiatry* 2, 618–24, 2015. Freeman D., Evans R., Lister R., Antley A. & Dunn G. 'Height, Social Comparison, and Paranoia: An Immersive Virtual Reality Experimental Study'. *Psychiatry Research* 218, 348–52, 2014. Kirkbride J. B., Errazuri A., Croudace T. J., Morgan C., Jackson D., Boydell J., Murray R. M. & Jones P. B. 'Incidence of Schizophrenia and Other Psychoses in England, 1950–2009: A Systematic Review and Meta-Analyses'. *PLOS ONE* 7, e31660, 2012.

p. 212 *Race and the prevalence of psychosis* – Schofield P., Ashworth M. & Jones R. 'Ethnic Isolation and Psychosis: Re-Examining the Ethnic Density Effect'. *Psychological Medicine* 41, 1263–9, 2011.

p. 212 *Exposure to social threat increases paranoid thinking* – Saalfeld V., Ramadan Z., Bell V. & Raihani N. J. 'Experimentally Induced Social Threat Increases Paranoid Thinking'. *Royal Society Open Science* 5, 180569, 2018.

p. 215 *more than half of all Americans endorse at least one conspiracy theory* – Oliver E. J. & Wood T. J. 'Conspiracy Theories and the Paranoid Style(s) of Mass Opinion'. *American Journal of Political Science* 58, 952–66, 2014.

p. 215 *A study measuring 12,000 letters written to the editors of the* New York Times *and the* Chicago Tribune – Uscinski J. E. & Parent J. M. *American Conspiracy Theories*. Oxford University Press, 2014.

p. 217 *Delusional beliefs are strange because no one else endorses them* – Bell V., Raihani N. J. & Wilkinson S. 'De-Rationalising Delusions'. *Clinical Psychological Science*, 2021. https://doi.org/10.1177/2167702620951553.

p. 217 *Beliefs as signals of group membership* – Williams D. 'Socially Adaptive Belief'. *Mind & Language*, 1–22, 2020.

p. 218 *Our brains seem to be chock-full of software that enables us to defend beliefs* – Mercier H., & Sperber D. 'Why do humans reason? Arguments for an argumentative theory'. *Behavioral and Brain Sciences, 34*, 57–111.

p. 218 *How ideology shapes cognition* – Van Bavel J. & Pereira A. 'The Partisan Brain: An Identity-Based Model of Political Belief'. *Trends in Cognitive Sciences* 22, 213–24, 2018.

p. 219 *US-based study measuring partisan differences in physical distancing* – Gollwitzer A., Martel C., Brady W. J., Pärnamets P., Freedman I. G., Knowles E. D. & Van Bavel J. J. 'Partisan differences in physical distancing are linked to health outcomes during the COVID-19 pandemic'. *Nature Human Behaviour*, 1–12, 2020.

17. TAKE BACK CONTROL

p. 222 *Reproductive skew is low among human males, relative to other primate species* – von Rueden C. R. & Jaeggi A. V. 'Men's Status and Reproductive Success in 33 Nonindustrial Societies: Effects of Subsistence, Marriage System, and Reproductive Strategy'. *Proceedings of the National Academy of Sciences* 113, 10824–9, 2016.

p. 223 *'Of course we have headmen. Every man is headman of himself.'* – Boehm C. *Hierarchy in the Forest: The Evolution of Egalitarian Behavior*. New edition. Harvard University Press, 2001.

p. 223 *Status is achieved via respect and prestige* – von Rueden C. 'Making and Unmaking Egalitarianism in Small-Scale Human Societies'. *Current Opinion in Psychology*, 33, 167–71, 2020. Cheng J. T., Tracy J. L., Foulsham T., Kingstone A. & Henrich J. 'Two Ways to the Top: Evidence That Dominance and Prestige Are Distinct yet Viable Avenues to Social Rank and Influence'. *Journal of Personality and Social Psychology* 104, 103–25, 2013.

p. 224 *Kalina passage* – Gillin J. 'Crime and punishment among the Barama River Carib of British Guiana'. *American Anthropologist* 36, 331–44, 1934.

p. 227 *Leadership in egalitarian hunter-gatherer societies* – Garfield Z. H., von Rueden C. & Hagen E. H. 'The Evolutionary Anthropology of Political Leadership'. *The Leadership Quarterly* 30, 59–80, 2019.

p. 228 *A theoretical model showing how preferences for hierarchy could emerge* – Powers S. T. & Lehmann L. 'An Evolutionary Model Explaining the Neolithic Transition from Egalitarianism to Leadership and Despotism'. *Proceedings of the Royal Society B: Biological Sciences* 281, 20141349, 2014.

p. 230 *The ritualised killing of humans became common* – Watts J., Sheehan O., Atkinson Q. D., Bulbulia J. & Gray R. D. 'Ritual Human Sacrifice Promoted and Sustained the Evolution of Stratified Societies'. *Nature* 532, 228–31, 2016.

p. 232 *Lanchester's Square Law* – Johnson D. P. & MacKay N. J. 'Fight the Power: Lanchester's Laws of Combat in Human Evolution'. *Evolution and Human Behavior* 36, 152–63, 2015.

p. 232 *Extremely small frequency of rebellion aboard slave ships* – Marcum A. & Skarbek D. 'Why Didn't Slaves Revolt More Often during the Middle Passage?'. *Rationality and Society* 26, 236–62, 2014.

18. VICTIMS OF COOPERATION

p. 235 *Uber drivers switching off app to generate surge pricing* – 'Uber, Lyft Drivers Manipulate Fares at Reagan National Causing Artificial Price Surges'. *WJLA*, 16 May 2019. http://wjla.com/news/local/uber-and-lyft-drivers-fares-at-reagan-national.

p. 236 *Corruption, bribery and nepotism as forms of cooperation* – Greif A. & Tabellini G. 'The Clan and the Corporation: Sustaining Cooperation in China and Europe'. *Journal of Comparative Economics* 45, 1–35, 2017. Muthukrishna M. 'Corruption, Cooperation, and the Evolution of Prosocial Institutions'. *SSRN Electronic Journal*, 2017. https://doi.org/10.2139/ssrn.3082315.

p. 237 *To put this another way, one might mistrust someone who 'always helps his friends'. But a similarly damning accusation can be levelled at someone who 'doesn't even help their friends'* – This is a mash-up of a quotation from the following source: Hampden-Turner C. & Trompenaars F. *Riding the Waves of Culture: Understanding Diversity in Global Business*. Hachette UK, 2011.

p. 237 *Smaller circle of moral regard among political conservatives* – Waytz A., Iyer R., Young L., Haidt J. & Graham J. 'Ideological Differences in the Expanse of the Moral Circle'. *Nature Communications* 10, 1–12, 2019.

p. 239 *Political ideology and scope of concern about impacts of COVID-19* – Raihani N. J. & de-Wit L. 'Factors Associated With Concern, Behaviour & Policy Support in Response to SARS-CoV-2, 2020'. https://doi.org/10.31234/osf.io/8jpzc.

p. 239 *Cross-cultural differences in scope of moral regard* – Yamagishi T., Jin N. & Miller A. S. 'In-Group Bias and Culture of Collectivism'. *Asian Journal of Social Psychology* 1, 315–28, 1998. Greif A. & Tabellini G. 'The Clan and the Corporation: Sustaining Cooperation in China and Europe'. *Journal of Comparative Economics* 45, 1–35, 2017. Jha C. & Panda B. 'Individualism and Corruption: A Cross-Country Analysis'. *Economic Papers: A Journal of Applied Economics and Policy* 36, 60–74, 2017.

p. 240 *Italy as an illustrative case-study* – Guiso L., Sapienza P. & Zingales L. 'Long-Term Persistence'. *Journal of the European Economic Association* 14, 1401–36, 2016. Reher D. S. 'Family Ties in Western Europe: Persistent Contrasts'. *Population and Development Review* 24, 203–34, 1998. Baldassarri D. 'Market Integration Accounts for Local Variation in Generalized Altruism in a Nationwide Lost-Letter Experiment'. *Proceedings of the National Academy of Sciences* 117, 2858–63, 2020.

p. 241 *The 2019, international lost-wallet experiment* – Cohn A., Maréchal M. A., Tannenbaum D. & Zünd C. L. 'Civic Honesty around the Globe'. *Science* 365, 70–3, 2019.

p. 242 *Material security* – Hruschka D. 'Parasites, Security, and Conflict: The Origins of Individualism and Collectivism'. *Evonomics*, 18 November 2015. https://evonomics.com/a-new-theory-that-explains-economic-individualism-and-collectivism/. Welzel C. *Freedom Rising: Human Empowerment and the Quest for Emancipation.* Cambridge University Press, 2013. Hruschka D. J. & Henrich J. 'Economic and Evolutionary Hypotheses for Cross-Population Variation in Parochialism'. *Frontiers in Human Neuroscience* 7, 559, 2013.

p. 244 *How material security varies across latitudes* – Van de Vliert E. & Van Lange P. A. M. 'Latitudinal Psychology: An Ecological Perspective on Creativity, Aggression, Happiness, and Beyond'. *Perspectives on Psychological Science* 14, 860–84, 2019.

p. 245 *Cooperation in the wake of COVID-19* – openDemocracy. 'The Social Support Networks Stepping up in Coronavirus-Stricken China'. 17 March 2020, https://www.opendemocracy.net/en/oureconomy/social-support-networks-springing-coronavirus-stricken-china/. 'Solidarity in Times of Corona in Belgium'. 24 March 2020, https://press.vub.ac.be/solidarity-in-times-of-corona-in-belgium. 'The Horror Films Got It Wrong. This Virus Has Turned Us into Caring Neighbours', *The Guardian* 31 March 2020, https://www.theguardian.com/commentisfree/2020/mar/31/virus-neighbours-covid-19.

p. 245 'Beers, Deer and Heroes: Heart-warming Moments in Coronavirus Britain'. *The Guardian*, 2 April 2020, https://www.theguardian.com/world/2020/apr/02/beers-deer-heroes-heartwarming-moments-coronavirus.

p. 245 *What we saw in the wake of this crisis, however, was a resurgence of cooperation – The bookstore in my neighbourhood* – is called Review and is on Bellenden Road in Peckham Rye. They're brilliant – support your local bookstore!

p. 246 *Self-interest in the wake of COVID-19* – 'New York's Andrew Cuomo Decries "eBay"-Style Bidding War for Ventilators'. *The Guardian*, 31 March 2020, https://www.theguardian.com/us-news/2020/mar/31/new-york-andrew-cuomo-coronavirus-ventilators.

p. 247 *in January 2019, a rare bluefin tuna weighing more than 270 kilos sold for $3.1 million* – 'Tuna Sells for Record $3 Million in Auction at Tokyo's New Fish Market', CNBC, 5 January 2019. https://www.cnbc.com/2019/01/05/tuna-sells-for-record-3-million-in-auction-at-tokyos-new-fish-market.html.

p. 250 *'think global, act local'* – Ostrom E., Burger J., Field C. B., Norgaard R. B. & Policansky D. 'Revisiting the Commons: Local Lessons, Global Challenges'. *Science* 284, 278, 1999.

p. 250 *Catch-share systems are effective so long as they are perceived to be legitimate* – Turner R. A., Addison J., Arias A, Bergseth B. J., Marshall N. A., Morrison T. H. & Tobin R. C. 'Trust, Confidence, and Equity Affect the Legitimacy of Natural Resource Governance'. *Ecology and Society* 21, 18, 2016.

p. 250 *'We Are Still In'* – https://www.wearestillin.com/.

p. 252 *Reduced coal consumption in China during COVID-19 pandemic* – Carbon Brief. 'Analysis: Coronavirus Temporarily Reduced China's CO2 Emissions by a Quarter', 19 February 2020. https://www.carbonbrief.org/analysis-coronavirus-has-temporarily-reduced-chinas-co2-emissions-by-a-quarter.

p. 252 *The success of Swiss e-bike trials* – Moser C., Blumer Y. & Hille S. L. 'E-Bike Trials' Potential to Promote Sustained Changes in Car Owners Mobility Habits'. *Environmental Research Letters* 13, 044025, 2018.

p. 252 *Changing commuters' habits* – Larcom S., Rauch F. & Willems T. 'The Benefits of Forced Experimentation: Striking Evidence from the London Underground Network'. *The Quarterly Journal of Economics* 132, 2019–55, 2017.

p. 253 *'acknowledged descendant of an Ape'* – Charles Darwin in correspondence to Frances Julia Wedgwood, July 1861. https://www.darwinproject.ac.uk/letter/?docId=letters/DCP-LETT-3206.xml. The full sentence reads: 'I admire the beautiful scenery more than could be reasonably expected of an acknowledged descendant of an Ape.'

INDEX